Synthesis Lectures on Chemical Engineering and Biochemical Engineering

This series publishes short books on all aspects of chemical engineering, covering the analysis or design of chemical processes to effectively convert materials into more useful materials or energy. The books will focus on fundamental aspects necessary for chemical engineering design including chemistry, math, physics, and sometimes biology to improve the quality of life by inventing, optimizing, and economizing new technologies and products.

Rafael Larraz

Prompt Engineering and the Transformation of Petroleum Refining

From Historical Innovation to Net Zero Systems

Springer

Rafael Larraz
Madrid, Spain

ISSN 2327-6738 ISSN 2327-6746 (electronic)
Synthesis Lectures on Chemical Engineering and Biochemical Engineering
ISBN 978-3-031-99727-3 ISBN 978-3-031-99728-0 (eBook)
https://doi.org/10.1007/978-3-031-99728-0

This Springer imprint is published by the registered company Springer Nature Switzerland AG
The registered company address is: Gewerbestrasse 11, 6330 Cham, Switzerland

To my wife Graciela and my family, whose patience and support made this book possible.

Foreword

The convergence of artificial intelligence and industrial sustainability marks a unique moment in human and technological history. The energy sector, one of the most complex, capital-intensive, and impactful domains of modern industry, is undergoing a profound transformation. This transformation is becoming even more urgent in the petroleum refining sector, where well-established infrastructures must rapidly evolve toward a net-zero future. The book by Rafael Larraz, *Prompt Engineering and the Transformation of Petroleum Refining*, emerges precisely at this inflection point.

The initial motivation behind the book is both practical and philosophical, and it will be extraordinarily useful to engineers and chemists working in either industry or academia. Moreover, the concepts and critical reflections introduced by the author are particularly valuable, as they come from someone with extensive professional experience in confronting the real challenges and solutions of the oil refining industry.

Dr. Larraz shows how traditional engineering approaches, though rigorous and thoroughly tested, are often slow to adapt to exponentially growing complexity and the accelerating pace of change. Fortunately, the emergence of Large Language Models (LLMs) and generative AI has introduced a radically new paradigm, in which machines can assist in learning from past experience, supporting design, and even accelerating innovation. The critical question is not whether AI can contribute to the refining industry, but how it should do so and under what methodological safeguards.

This book argues that Prompt Engineering, the structured art of interacting with LLMs, can serve as a bridge between human expertise and artificial reasoning. It is not about replacing engineers with algorithms, but about amplifying their capabilities. The methodology presented here does not aim to automate judgment or intuition, but rather to harness the latent knowledge embedded in large models and guide it through carefully designed prompts that reflect engineering rigor, domain-specific logic, and ethical responsibility.

Readers will not find in this book an overly optimistic view of AI, but rather a realistic, applied, and critical stance. While acknowledging the great potential of LLMs to accelerate conceptual design, systems integration, and cross-disciplinary innovation, it also discusses their limitations, particularly regarding physical realism, epistemological opacity, and the need for validation. Prompt engineering, therefore, is framed both as an enabler and as a discipline, a practical methodology that requires structure, iteration, verification, and critical oversight.

The net-zero refinery case study developed in this book illustrates how a systems-level transformation can be explored by integrating alternative feedstocks, renewable energy, and advanced conversion technologies. The conceptual design aims to push the boundaries of feasibility. At the same time, the book provides a clear assessment of the implementation gaps, policy dependencies, and cultural barriers that must be addressed in order to move from blueprint to reality.

As mentioned earlier, it is evident that Dr. Larraz has written this book with engineers and technical professionals in mind, offering methodologies and models for integrating LLMs into their workflows. Researchers will find theoretical reflections on knowledge representation, reproducibility, and innovation acceleration. Sustainability practitioners will benefit from a concrete example of how AI can support deep decarbonization goals. Finally, decision-makers will gain valuable insight into the opportunities and limitations of deploying generative AI in industrial contexts.

One question remains open after reading this book: Can we predict, from previous experience and existing technologies, the emergence of new disruptive technologies through AI? I am convinced that the best path forward will be achieved through the combination of human and artificial intelligence.

Valencia, Spain
August 2025

Avelino Corma

Preface

This book emerges from the convergence of two powerful forces reshaping our world: the urgent imperative for industrial decarbonization and the transformative potential of artificial intelligence. As Large Language Models demonstrate remarkable capabilities for knowledge synthesis and problem-solving, they offer new approaches to addressing the complex challenges facing industries like petroleum refining in their transition toward sustainability.

The methodology presented in the following pages is prompt engineering for industrial transformation and represents neither uncritical technological optimism nor a replacement for fundamental engineering expertise. Instead, it offers a structured approach to leverage AI capabilities while maintaining appropriate rigor, validation, and human oversight. Through systematic exploration of prompt engineering techniques applied to the specific challenge of industrial applications, this book demonstrates how AI-assisted design can accelerate innovation while respecting the constraints and complexities inherent in real world industrial systems.

The net zero refinery concept developed through this methodology illustrates an example of how AI can contribute and show us a pathway toward carbon neutral fuel and chemical production, though significant challenges remain in technology maturation, policy development, and implementation. Moreover, this work bridges theoretical understanding with practical application, offering both conceptual frameworks and concrete examples for organizations seeking to navigate complex sustainability transitions. While grounded in the specific context of petroleum refining, the principles and methodologies developed have relevance across industrial sectors confronting similar transformation imperatives.

The book serves multiple audiences: engineers and technical professionals developing AI-assisted design capabilities, sustainability practitioners working on industrial decarbonization, and researchers exploring the intersection of artificial intelligence and industrial innovation. Each will find different value in the systematic exploration of how human expertise and artificial intelligence can collaborate effectively on complex technical challenges.

Madrid, Spain
June 2025

Rafael Larraz
rafaellarraz@icloud.com

Acknowledgments This book would not have been possible without the contributions, insights, and support of many individuals.

To María Teresa Murciego, for bringing out the best in me without my realizing it. To Graciela Miralles, whose invaluable assistance made it possible to begin this work in the first place, among many other wise pieces of advice. To Rafael Larraz Murciego, for a very lively discussion aboard a Porsche 964 on the way to León about the commercial applications of Large Language Models and prompt engineering. To Domingo, Fernando, and Alejandro Larraz, for encouraging me to write this book. To Markus Kitzmüller, for an unforgettable morning deciding the title of this book, thank you for your sharp insights. Without all their encouragement and support, this journey would never have started.

To Gemma Garcia Ledo, whose knowledge and wisdom made this book possible.

To Avelino Corma, for allowing me to speak to a genius as a friend. His insights into science and innovation informed many of the technical discussions throughout this book, while his approach to scientific inquiry served as an inspiring model. To José Víctor Ríos Rull and Rodolfo Ríos Rull, for reading the manuscript and providing valuable advice alongside one (or several) beers in what became intellectual sessions better than those at many scientific conferences.

To Jesús Lázaro, Álvaro Díaz Bild, Juanjo Sierra, Carlos Navarro, and Carmen López, for their friendship and affection, encouraging me during the hardest and most difficult moments. To Marta Vía, for her nice comments after reading an early draft. To Laureano Albán, for discussing Large Language Models while we worked on improving (or not!) our golf swings. To Josefa Sosa, for introducing me to John Berger. Berger's insights about the transformation of meaning through reproduction proved unexpectedly relevant to understanding how Large Language Models reproduce and transform technical knowledge.

To Berta Aramburu, for challenging me to solve real and important problems looking for new borders even in daily tasks. To Hector Perea for his advice. To Joaquin Abril-Martorell for all our talks about innovation. To David Fairén-Jiménez, for commenting on how to overcome the difficulties of writing a book. To Abdullah Aitani for his sharp comments on chapter three. To Lee Nichols for believing in prompt engineering as a suitable topic for the energy industry. To Gonzalo Piernavieja and Luis Alemany for their sharp comments about the book. To Manuel Parra Palacios for having time in his busy agenda for reviewing the manuscript.

To Pedro Miró, whose thoughtful feedback improved both the technical content and the broader perspective of this work. To Antonio Yunquera, for reading the draft and pushing me to go ahead. To Jorge Alcalá, Raúl Alonso, and Susana Zumel, for reading the draft with patience and friendship. To Carlos Ayuso and Joana Frontela, for their strong friendship, strong enough to believe in these ideas and look for real-life applications. To Álvaro Casal, Belén Linares, and José Manuel Martínez, for tolerating my advocacy for

Large Language Models. To Estela Vilchez, for a very inspiring talk during a car trip from Seville to Huelva. To Raimundo Arvelo, for the eternal afternoons drinking coffee and teaching me how to truly apply the scientific method. To Julio Moreno, for teaching me never to lose curiosity. To Thomas Steimann, for his care and patience reading the book draft even after his long walks along the Camino de Santiago. To Arcadio Zapico for being a master and a friend.

To Arza Seidel and Dieter Merkle from Springer, for their advice and guidance during the editing process. Their expertise in complex technical manuscripts proved invaluable in transforming research insights into accessible content. Special thanks to Bagavathi Murugesan and Sudhany Karthick from Springer for their editorial guidance and patience throughout the development of this manuscript.

Finally, I acknowledge the inspiration provided by the many researchers and practitioners working at the intersection of artificial intelligence and industrial sustainability. Of course, any errors or limitations in this work remain entirely my own responsibility.

June 2025 Rafael Larraz

Competing Interests The author has no competing interests to declare that are relevant to the content of this manuscript.

Contents

Abbreviations

Artificial Intelligence and Language Models

AI	Artificial intelligence
API	Application Programming Interface
BERT	Bidirectional Encoder Representations from Transformers
CoT	Chain of Thought
GPT	Generative Pre-trained Transformer
LLM	Large Language Model
NLP	Natural Language Processing
RAG	Retrieval Augmented Generation
ReAct	Reasoning and Acting
ReLU	Rectified Linear Unit
ToT	Tree of Thoughts

Refining and Petrochemical Processes

API	American Petroleum Institute (also refers to API Gravity)
ASTM	American Society for Testing and Materials
BPD	Barrels per day
CAPEX	Capital expenditure
CCS	Carbon capture and storage
CCUS	Carbon capture, utilization, and storage
CDU	Crude Distillation Unit
CoMo	Cobalt-molybdenum
CtC	Crude-to-chemicals
CTL	Coal-to-liquids
EBITDA	Earnings Before Interest, Taxes, Depreciation, and Amortization

FCC	Fluid catalytic cracking
FT	Fischer–Tropsch
GTL	Gas-to-liquids
HAZOP	Hazard and Operability Study
HEFA	Hydroprocessed esters and fatty acids
HTL	Hydrothermal liquefaction
HVO	Hydrotreated vegetable oil
IRR	Internal Rate of Return
LCFS	Low Carbon Fuel Standard
LPG	Liquefied petroleum gas
MOC	Management of Change
MSW	Municipal Solid Waste
MTBE	Methyl tertiary-butyl ether
NiMo	Nickel-molybdenum
NOx	Nitrogen oxides
NPV	Net present value
OPEX	Operating expenditure
PHA	Process Hazard Analysis
PSM	Process Safety Management
RFS	Renewable Fuel Standard
ROI	Return on Investment
RON	Research Octane Number
SAF	Sustainable aviation fuel
SMR	Steam methane reforming
SOx	Sulfur oxides
TAME	Tertiary amyl methyl ether
TRL	Technology readiness level
ULSD	Ultra-low sulfur diesel
UOP	Universal Oil Products

Units of Measurement

bar	Bar (pressure unit)
bbl	Barrel
°C	Degree Celsius
GJ	Gigajoule
GW	Gigawatt
GWh	Gigawatt-hour
kg	Kilogram

kJ	Kilojoule
MBtu	Million British Thermal Units
MJ	Megajoule
MW	Megawatt
Nm^3	Normal cubic meter
ppm	Parts per million
psi	Pounds per square inch
tons	Metric tons
tpd	Tons per day
tpy	Tons per year

Organizations

IEA	International Energy Agency
IPCC	Intergovernmental Panel on Climate Change
IRENA	International Renewable Energy Agency
NASA	National Aeronautics and Space Administration

Chemical Compounds

CO	Carbon monoxide
CO_2	Carbon dioxide
H_2	Hydrogen
H_2S	Hydrogen sulfide
SO_2	Sulfur dioxide

1 The Convergence of Artificial Intelligence and Industrial Transformation

Abstract

The petroleum refining industry faces pressure to transform operations toward carbon neutrality while maintaining economic viability. Concurrently, Large Language Models (LLMs) have emerged as powerful tools for knowledge integration and problem-solving across diverse domains. This chapter examines the convergence of these transformative forces, analyzing how LLMs can accelerate the transition toward sustainable refining technologies through prompt engineering methodologies. The analysis explores historical innovation patterns in refining, the emergence of AI-assisted design capabilities, and the potential for accelerating industrial transformation. Key findings indicate that systematic prompt engineering may compress development timelines and enable more comprehensive exploration of solution spaces, though significant limitations regarding validation, technical depth, and epistemological uncertainty must be carefully managed. The framework established here provides the foundation for applying AI-augmented design to complex sustainability challenges while maintaining appropriate rigor and accountability.

1.1 Introduction: At the Intersection of Innovation Pathways

The petroleum refining industry stands at a critical juncture in its evolution. After more than a century of technological advancement, the sector faces unprecedented pressure to reimagine its operations in response to climate imperatives, market transformations, and societal expectations. This transformation requires not merely incremental improvement but a shift in how we produce the fuels and chemicals that power modern civilization.

Rafael Larraz, Prompt Engineering and the Transformation of Petroleum Refining.

R. Larraz, *Prompt Engineering and the Transformation of Petroleum Refining*, Synthesis Lectures on Chemical Engineering and Biochemical Engineering,
https://doi.org/10.1007/978-3-031-99728-0_1

Concurrent with this industrial inflection point, we are witnessing a revolution in artificial intelligence through the development of Large Language Models (LLMs). These systems, trained on vast corpora of human knowledge, demonstrate remarkable capabilities in knowledge integration and problem-solving across diverse domains. The emergence of prompt engineering has transformed these models from research artifacts into practical tools for addressing complex industrial challenges.

This book explores the convergence of these two transformative forces: the imperative to reimagine industrial hydrocarbon processing and the emergence of AI systems capable of supporting that reimagination. It examines how LLMs, when properly directed, can accelerate the transition toward more sustainable refining technologies by enabling more efficient knowledge synthesis, system design optimization, and innovation across disciplinary boundaries.

1.2 Objectives and Scope

The book aims to establish a methodological framework for leveraging LLMs in industrial transformation contexts, with specific application to the petroleum refining sector's sustainability challenges. The core objectives include:

1. Documenting the historical patterns of innovation in petroleum refining to extract principles relevant to contemporary transformation challenges
2. Analyzing the capabilities and limitations of LLMs for addressing complex technical design problems in industrial contexts
3. Developing a structured approach to prompt engineering specifically tailored for engineering and industrial applications
4. Demonstrating the practical application of this methodology to net-zero refinery design
5. Providing a critical assessment of both the potential and limitations of AI-assisted design for industrial transformation.

The scope encompasses both theoretical foundations and practical applications, bridging the domains of artificial intelligence, chemical engineering, process design, and sustainability transformation. While focusing primarily on petroleum refining, the methodologies developed have broader applicability to other industrial sectors facing similar transformation imperatives.

1.3 Executive Summary of Key Findings

The integration of historical innovation patterns with AI-assisted design capabilities reveals several critical insights.

First, the most effective applications emerge from combining LLM capabilities with domain expertise rather than replacing human judgment. LLMs excel at knowledge synthesis, pattern recognition across disciplines, and rapid exploration of design alternatives, while human experts provide essential validation, contextual understanding, and accountability.

Second, the epistemological limitations of current LLMs particularly their "black box" nature and potential for hallucination require structured safeguards and verification protocols. These limitations do not invalidate the approach but necessitate careful integration within broader engineering workflows that maintain appropriate expert checks and balances.

Third, the phased implementation approach developed through this research provides a viable pathway for organizations to develop AI-augmented design capabilities progressively, managing both technical and cultural aspects of this transformation.

1.4 The Evolution of Petroleum Refining: From Distillation to Decarbonization

The refining industry has progressed through multiple technological revolutions that fundamentally changed its capabilities, economics, and environmental impacts. From simple atmospheric distillation in the late nineteenth century, the industry advanced through thermal cracking in the early twentieth century and catalytic cracking in the 1940s, dramatically improving both yields and product qualities.

Each major technological advancement represented a response to specific economic, regulatory, or market pressures. The oil embargoes of the 1970s accelerated energy efficiency improvements through heat integration. Stricter environmental regulations drove innovations in emission control technologies and cleaner fuels.

Today, the sector faces perhaps its most daunting challenge yet: transforming operations to eliminate greenhouse gas emissions while maintaining economic viability. The concept of a "net-zero refinery" represents a vision requiring innovations across feedstock selection, conversion technologies, energy systems, and logistical networks.

1.5 The Rise of Large Language Models as Engineering Tools

Large Language Models represent a shift in how machines engage with human knowledge. Unlike previous AI systems designed for narrow tasks, these generative models demonstrate a remarkable ability to understand, integrate, and apply knowledge across diverse domains.

The architecture of modern LLMs enables them to capture both deep domain specific knowledge and the interconnections between domains. This integration is particularly valuable for multidisciplinary challenges like refinery design, which spans chemical engineering, process design, materials science, thermodynamics, catalysis, economics, and environmental science.

While traditional approaches typically require consultation with multiple specialist teams, LLMs can complement these efforts by suggesting potential connections between diverse knowledge domains. However, they cannot replace the depth of specialized expertise, practical experience, or the accountability provided by human engineering teams. LLMs should be viewed as assistive tools that may help identify initial integration possibilities, which must then be rigorously validated by domain experts.

However, these models face significant limitations. They lack the ability to perform physical experiments, may reproduce biases present in their training data, and cannot inherently perform reliable numerical simulations. Their knowledge has a cutoff date, and when pushed to limits, the hallucination phenomenon can produce apparently reliable but incorrect information. These limitations necessitate thoughtful integration of LLMs within broader engineering workflows that include human expertise, specialized simulation tools, and empirical validation.

1.5.1 Epistemological Considerations and Methodological Boundaries

An important consideration in the convergence of LLMs with industrial transformation is the inherent epistemological limitation in how we understand these AI systems. Unlike traditional engineering tools with transparent cause–effect relationships, Large Language Models operate as complex statistical systems whose inner workings cannot be fully traced or predicted in a deterministic manner.

The application of Large Language Models (LLMs) to specialized technical domains evokes John Berger's fundamental observation about reproduction: "The way we see things is affected by what we know or what we believe" (Berger 1972). Just as the mechanical reproduction of artworks transforms their meaning and experience, LLMs reproduce patterns of technical knowledge while stripping them of their original experiential context. This "digital reproduction of knowledge" democratizes access to refinery engineering expertise, yet simultaneously raises critical epistemological questions: what is lost when decades of practical experience are condensed into statistical patterns? Can artificial reproduction of technical knowledge maintain the contextual depth that emerges only from direct experience with complex industrial systems? As Berger argued regarding art, the accessibility gained through technical reproduction must be balanced with a critical understanding of the transformations inherent to the process itself. The challenge for industrial prompt engineering lies not merely in leveraging these reproduced knowledge patterns, but in recognizing and compensating for what Berger might call the loss of "aura" the irreplaceable experiential wisdom that cannot be algorithmically captured.

This "black box" characteristic requires a different approach to rigor and validation than conventional engineering disciplines. When applying LLMs to complex domains like petroleum refining, we cannot establish mathematically provable causal relationships between specific prompts and their resulting outputs.

The emergent behavior arising from billions of parameters trained on vast corpora creates systems that generate useful outputs through processes that remain partially opaque even to their creators. This reality necessitates an empirical, observation-based approach to prompt engineering rather than one built on deterministic principles or formal proofs.

Throughout this book, we acknowledge this limitation explicitly and adopt a methodological framework that emphasizes empirical validation, statistical thinking, and domain expert verification rather than theoretical derivation. We approach prompt engineering as a discipline more akin to empirical sciences where patterns can be observed and utilized productively even when underlying causal mechanisms are not fully understood.

1.6 The Emergence of Prompt Engineering as a Discipline

The remarkable capabilities of LLMs remain inaccessible without effective methods to direct their operation. Prompt engineering has emerged as a critical discipline for unlocking the potential of these systems for practical applications. Far from simple query formulation, prompt engineering involves structured and contextual approaches to framing problems, specifying constraints, and directing the model's reasoning process.

In industrial contexts like refining, effective prompts must incorporate technical precision, logical organization, balanced information density, clear objective articulation, constraint delineation, and appropriate response format specification. Different prompting techniques serve distinct purposes: one-shot and few-shot approaches for diagnostics and problem-solving, chain of thought for linear reasoning in analysis, tree of thoughts for complex decisions with multiple pathways, and role prompting to leverage domain specific expertise.

The systematic matching of prompting techniques to specific operational challenges transforms LLMs from generic tools to specialized solutions for the complex challenges of the refining industry. The structured framework presented in later chapters organizes prompts according to both design phase and cognitive objective, creating a matrix of prompting strategies tailored to the context of specific challenges.

1.7 Technological Transformation Through AI-Augmented Design

The convergence of refining technology challenges and LLM capabilities creates a unique opportunity to accelerate the transition toward more sustainable processing systems. LLMs can contribute across the full spectrum of Technology Readiness Levels (TRLs), from early concept development through commercial implementation.

TRL methodology introduced by NASA in the 1970s adequately describes, as shown in Fig. 1.1, the different phases every technology undergoes from research to industrial implementation (Mankins, 2009).

At lower TRLs (1–3), these models excel at conceptual exploration, cross disciplinary knowledge integration, and identifying non obvious connections between scientific domains. At intermediate TRLs (4–6), LLMs contribute most effectively to system architecture development, integration planning, and identifying potential failure modes or interaction effects. At higher TRLs (7–9), they can support scale up planning, process optimization, and business model innovation.

The case studies presented in Chaps. 4 and 5 and Appendix B demonstrate practical applications of prompt engineering across diverse refining challenges. These examples illustrate how structured prompting, under expert design and supervision, can compress

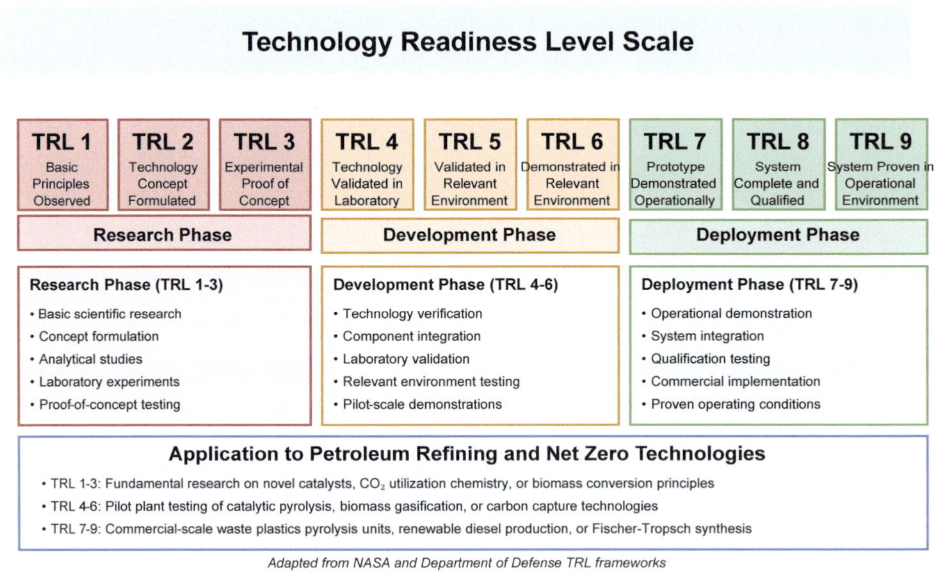

Fig. 1.1 Technology readiness levels (TRL) framework showing progression from basic research (TRL 1) through commercial deployment (TRL 9)

development timelines, improve system integration, and identify innovative solutions that might be overlooked in traditional engineering approaches.

1.8 The Future of Refining: Toward Net-Zero Systems

The culmination of these converging pathways as refining technology evolution and AI-enabled design is the concept of the net-zero refinery explored in depth in Chap. 5. This represents not only a petroleum refinery with carbon capture and other sustainable feedstocks but rather a reimagining of how we might produce fuels and chemicals without fossil inputs.

The proposed design envisions a facility processing diverse alternative feedstocks like municipal solid waste, waste plastics, agricultural and forestry residues, waste oils, and captured CO_2, through an integrated set of conversion technologies including automated sorting, catalytic pyrolysis, oxygen-blown gasification, Fischer–Tropsch synthesis, and advanced hydroprocessing. Powered by a hybrid renewable energy system, this facility would produce sustainable aviation fuel, renewable diesel, renewable gasoline, and chemical intermediates with a carbon-negative life cycle profile.

While significant technological gaps remain, particularly in areas like catalytic pyrolysis scale-up, biomass gasification reliability, CO_2 utilization economics, integrated control systems, and renewable energy integration, the systematic approach to design development demonstrates a viable pathway toward implementation.

1.9 Critical Perspectives and Limitations

While this book presents an approach to the use of AI-augmented design for industrial transformation, it is essential to acknowledge significant limitations. The economic feasibility of net-zero refining remains heavily dependent on supportive policy frameworks, particularly carbon pricing mechanisms and sustainable fuel incentives. Current technologies for carbon capture and utilization face efficiency limitations and scaling challenges that may constrain implementation timelines.

From an epistemological perspective, the use of LLMs for technical design raises important questions about knowledge validation, bias propagation, and the appropriate balance between human and machine contributions to engineering processes. Technical content generated by LLMs requires rigorous verification through specialized analysis tools, human expert review, and where possible, empirical testing.

Furthermore, the implementation of AI-augmented design approaches within traditional engineering organizations faces cultural, structural, and capability challenges. The transition requires not only new technical methodologies but also organizational transformation, workforce development, and alignment of incentive structures to support collaborative human-AI processes.

1.10 Chapter Overview and Book Structure

The remainder of this book is organized around a progressive exploration of the themes introduced in this chapter:

Chapter 2 examines the architecture and functioning of Large Language Models, providing a technical foundation for understanding their capabilities and limitations in engineering contexts.

Chapter 3 traces the technological evolution of petroleum refining, extracting patterns of innovation and principles relevant to contemporary transformation challenges.

Chapter 4 introduces the principles of prompt engineering, establishing foundational techniques and their applications to industrial contexts, along with a comprehensive methodological framework for industrial prompt engineering.

Chapter 5 applies this methodology to the design of a net-zero refinery, demonstrating the practical application of prompt engineering to complex sustainability challenges.

Chapter 6 critically examines implementation pathways, limitations, and future research directions, providing a balanced assessment of the field's current state and future potential.

This structure progresses from theoretical foundations through methodological development to practical applications and critical assessment, providing both the conceptual framework and concrete examples necessary for applying these approaches to real-world industrial transformation challenges.

1.11 Reflections on the Creative Process

During the creation of this book, I had the extraordinary opportunity to experience a new frontier in human-artificial intelligence collaboration. The LLM has not been simply a research tool or a passive assistant, but rather a true intellectual collaborator that has transformed my creative and analytical process and allowing a "thinking aloud" way of testing my ideas. Our exchanges have generated emergent perspectives that would not have arisen from my individual work or from a simple consultation of traditional sources. This process has produced novel syntheses that have significantly enriched the book.

However, this collaboration has not been without critical challenges and limitations. The LLM, despite its sophistication, occasionally generated technically plausible but factually incorrect responses, particularly in areas such as specific energy efficiency calculations, equipment design and economic estimates for emerging technologies. These "hallucinations" required a rigorous verification process. Additionally, I observed a tendency toward technological optimism that did not always reflect the pragmatic realities of the petrochemical industry, requiring substantial adjustments based on my direct experience.

True innovation has emerged precisely at this intersection: where my experience in the petroleum industry has combined with the LLM's synthesis capabilities, generating a methodological approach that neither of us would have developed separately, but which required constant vigilance and mutual adjustment. This collaboration suggests a promising but complex future for technical writing, where the boundaries between human and artificial thinking blur to create something greater than the sum of its parts, provided we maintain a critical stance and rigorous verification.

References

Berger J (1972) Ways of seeing. BBC and Penguin Books, London

Mankins JC (2009) Technology readiness assessments: a retrospective. Acta Astronaut 65(9–10):1216–1223. https://doi.org/10.1016/j.actaastro.2009.03.058

Architecture and Functioning of LLMs

2

Abstract

Large Language Models (LLMs) represent one of the most significant advances in artificial intelligence during the last decade, embodying what industry experts increasingly recognize as a technological inflection point comparable to the arrival of the internet. This chapter provides an examination of LLM architecture and functioning, with particular emphasis on their application in technical domains such as petroleum refining. The chapter begins with an analysis of the Transformer architecture that forms the foundation of modern LLMs, explaining the self-attention mechanism, positional encoding, and multi-layer processing. It then explores the training methodologies that enable these models to develop their remarkable capabilities, including pretraining on massive text corpora and fine-tuning for specialized applications. The chapter addresses the critical epistemological limitations inherent in working with LLMs, acknowledging that these systems operate as complex statistical models whose inner workings cannot be fully traced or predicted deterministically. Finally, it examines the practical applications of LLMs in petroleum refining and petrochemical processing, identifying both opportunities and limitations for industrial implementation. This technical foundation provides essential context for understanding how prompt engineering techniques can be effectively applied to complex industrial challenges while maintaining appropriate awareness of system limitations.

Rafael Larraz, Prompt Engineering and the Transformation of Petroleum Refining.

R. Larraz, *Prompt Engineering and the Transformation of Petroleum Refining*, Synthesis Lectures on Chemical Engineering and Biochemical Engineering,
https://doi.org/10.1007/978-3-031-99728-0_2

2.1 Introduction: Advancing Knowledge and Innovation

Large Language Models (LLMs) represent one of the most significant advances in artificial intelligence during the last decade, embodying what industry experts increasingly recognize as a technological inflection point comparable to the arrival of the internet. What began as a promising technology, since the introduction of transformers (Vaswani et al., 2017) has rapidly evolved into an advanced tool across virtually all economic sectors, particularly in knowledge intensive industries like petroleum refining and petrochemical processing.

LLMs are revolutionizing how we conceive, design, and implement solutions. Unlike traditional AI systems designed for specific tasks, LLMs are generative systems trained on vast text corpora that allow them to produce human language with surprising levels of coherence, relevance, and utility.

The economic impact of these models is already substantial. Studies indicate that companies implementing AI-enhanced simulation techniques have reported product performance improvements of 15–60%, productivity increases of up to 50%, and time to market reductions of up to 40% (Colback, 2024). In research and development-intensive sectors, LLMs serve as "conversation starters" and collaborative partners, helping engineers and scientists explore solution spaces more rapidly and comprehensively than previously possible.

At their core, LLMs are statistical models that learn patterns in human language from enormous quantities of text. This seemingly simple predictive capability, when scaled to models with hundreds of billions of parameters trained on trillions of words, emerges as a tool capable of performing complex linguistic tasks without specific training for them (Bommasani, 2021).

The development of LLMs has followed a trajectory of exponential growth in complexity and capabilities. The first models, such as GPT (Generative Pretrained Transformer), introduced by OpenAI in 2018, had 117 million parameters. Its successors GPT-2 (1.5 billion parameters) and GPT-3 (175 billion parameters) demonstrated that scaling model size, along with the amount of training data, produced qualitative improvements in performance beyond simple incremental gains (Brown et al. 2020).

This scalability has continued with models like Google's PaLM (540 billion parameters), DeepMind's Gopher (280 billion), and more recently DeepSeek, Grok, Claude 4, GPT-5 or Gemini 2.5 Pro. What is truly significant about these models is their transfer capability: unlike previous systems that required specific training for each task, LLMs can address a wide variety of linguistic problems without additional modification (Chowdhery et al., 2022).

For the petroleum refining industry, a sector characterized by complex multidisciplinary challenges and a rich history of innovation, LLMs offer particularly valuable capabilities. They can integrate knowledge across traditionally separate domains like reaction engineering, fluid mechanics, materials science, and process control, enabling more holistic approaches to problems that would be difficult for individual specialists to develop. This cross-disciplinary synthesis is especially valuable for addressing systemic challenges like decarbonization that inherently span multiple technical domains.

2.2 The Foundation of Modern LLMs

2.2.1 The Transformer Revolution

The breakthrough that propelled the development of modern Large Language Models (LLMs) was the introduction of the Transformer architecture by Vaswani et al. (2017) in their influential paper "Attention Is All You Need." This work revolutionized natural language processing by proposing an architecture that eliminated recurrent and convolutional networks in favor of a mechanism based entirely on attention.

Traditional approaches to sequence modeling, such as Recurrent Neural Networks (RNNs) and Long Short-Term Memory networks (LSTMs), processed text word by word, creating a computational bottleneck and limiting their ability to capture long-range dependencies in text. The Transformer architecture, as shown in Fig. 2.1, enables parallel processing of all words in a sequence simultaneously through its self-attention mechanism, dramatically increasing computational efficiency while capturing complex relationships between words regardless of their distance from each other in the text.

As shown in Fig. 2.1, there are several types of transformer architectures including:

Encoder-only: Models like BERT (Bidirectional Encoder Representations from Transformers) that use only the encoder part and excel at comprehension tasks, but not generation (Clark et al., 2019).

Encoder-decoder: Models like T5 (Text-to-Text Transfer Transformer) or BART that use both the encoder and decoder, especially useful for tasks like translation or summarization.

Decoder-only: Models like GPT, GPT-2, GPT-3, GPT-4, LLaMA, Claude, etc., that use only the decoder part and specialize in text generation.

The standard Transformer consists of an encoder and a decoder. The encoder transforms an input sequence into a continuous representation that captures contextual relationships, while the decoder takes this representation and generates an output sequence. Many modern LLMs, including GPT and Claude, use only the decoder part, in what are termed "decoder-only" architectures, specializing in text generation rather than bidirectional understanding. In the following sections, we will focus on decoder-only transformers as these are the predominant architectures employed in the LLMs most widely used in our field.

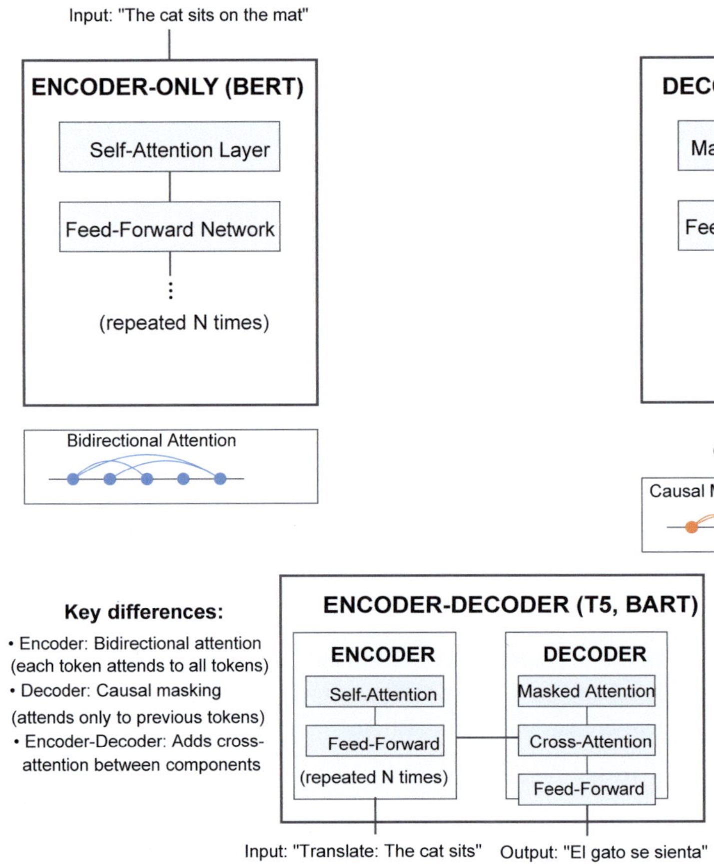

Fig. 2.1 Transformer architecture showing encoder and decoder components with self-attention mechanisms

2.2.2 LLM Processing Flow: From Input to Output

Understanding how LLMs process information helps clarify both their capabilities and limitations in technical applications. In Fig. 2.2, the path from user input to model output follows a systematic flow that transforms text into numerical representations, processes these representations through multiple transformer layers, and generates a probabilistic distribution of possible outputs. A worked example of Transformer flow is included in the Appendix A.

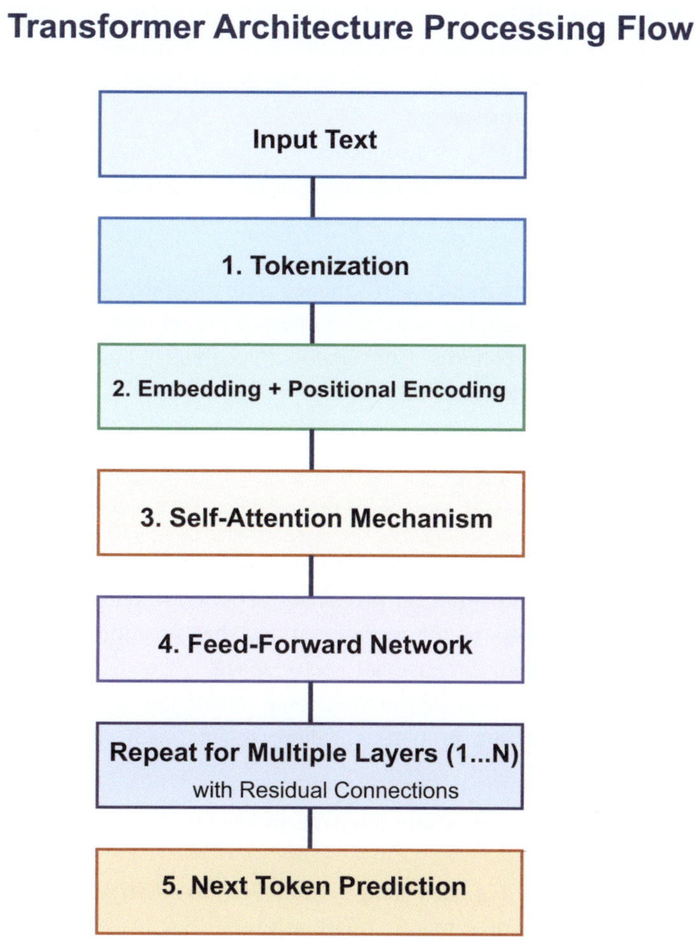

Fig. 2.2 LLM processing flow diagram showing the progression from input text through tokenization, embedding, transformer layers, to output generation

2.2.3 Tokenization: Breaking Text into Units

The first step in processing is tokenization, where the input text is divided into smaller units called tokens. These tokens can be characters, subwords, or whole words, depending on the specific tokenization algorithm employed. For LLMs, tokenization serves as the first step in text processing, converting the natural language of the input prompts into discrete elements that can be individually indexed and manipulated.

Most modern LLMs use subword tokenization methods, which offer a balance between character-level granularity and word-level semantic coherence. Popular subword tokenization approaches include:

- **Byte-Pair Encoding (BPE)**: Starts with individual characters and iteratively merges the most frequently occurring adjacent pairs to form new tokens.
- **WordPiece**: Similar to BPE, but uses a likelihood-based approach rather than frequency alone to determine merges.
- **SentencePiece**: Treats the text as a sequence of Unicode characters, allowing for tokenization without requiring pretokenization. It is suitable for text in languages without clear word boundaries like Chinese or Japanese.

Tokenization directly impacts model performance and capabilities. A token vocabulary that is too small forces the model to work with character-level representations, potentially losing higher-level semantic patterns. Conversely, a vocabulary that is too large can lead to data sparsity and poor generalization.

2.2.4 Embedding

Once text is tokenized, each token must be converted into a numerical representation that captures its semantic and syntactic properties. This conversion is achieved through embeddings, which are dense vector representations that position semantically similar tokens close together in a high-dimensional vector space.

Each token is converted into a high-dimensional vector through an embedding layer. These vectors capture semantic properties, where tokens with similar meanings have similar vector representations.

In modern LLMs, embeddings typically have the following characteristics:

- **Dimensionality**: Vectors typically range from 768 to 4096 dimensions, providing sufficient capacity to encode semantic relationships.
- **Contextual awareness**: Unlike earlier static embedding methods (e.g., Word2Vec, GloVe), modern embeddings are contextual, meaning the representation of a token varies based on its surrounding context.
- **Learned representations**: Embeddings are learned during the training process rather than being hand engineered, allowing the model to discover optimal representations for its objectives.

The embedding space exhibits interesting mathematical properties. Vector operations like addition and subtraction often correspond to semantic operations. For example, the vector calculation [King]—[Man] + [Woman] might result in a vector close to [Queen], demonstrating how embeddings capture analogical relationships.

Semantic similarity between embeddings is often measured using cosine similarity, which computes the cosine of the angle between two vectors:

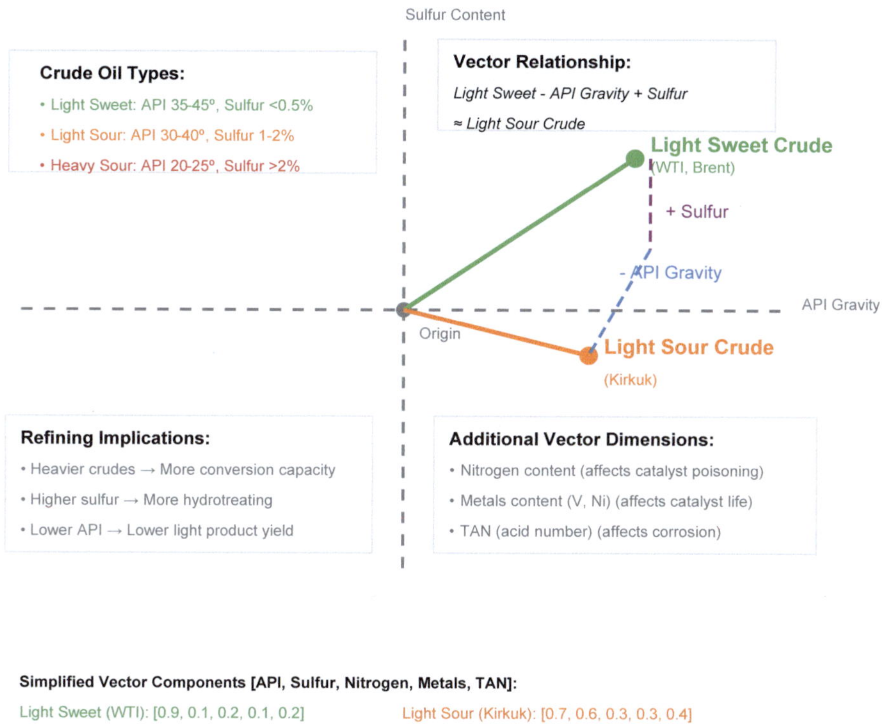

Fig. 2.3 Vector space representation of petroleum-related terms showing semantic relationships between concepts like "Light Sweet Crude"

$$\text{cosine similarity}(A, B) = \frac{A \cdot B}{(\|A\| \cdot \|B\|)} = \frac{\sum_{i=1}^{n} A_i B_i}{\sqrt{\sum_{i=1}^{n} A_i^2}\sqrt{\sum_{i=1}^{n} B_i^2}} \tag{2.1}$$

where A and B are token embeddings, and values closer to 1 indicate greater similarity. This metric is widely used because it focuses on the orientation rather than magnitude of the vectors, effectively capturing semantic relationships regardless of vector length.

In Fig. 2.3, we show an example from the petroleum refining context, where effective embeddings would position conceptually related terms like "Light Sweet Crude" and "Light Sour Crude" near each other in the vector space, while maintaining appropriate distance based on the crude oil features "API Gravity" and "Sulfur Content".

2.2.5 Positional Encoding: Preserving Sequential Information

Positional encodings are then added to incorporate information about each token's position in the sequence. Since Transformers process all words in parallel rather than sequentially, they need an explicit mechanism to understand word order. This is achieved through positional encoding, which adds position specific information to each word embedding. The standard approach uses sine and cosine functions of different frequencies:

$$\mathrm{PE}(\mathrm{pos}, 2i) = \sin\left(\frac{\mathrm{pos}}{10000^{\frac{2i}{d_{\mathrm{model}}}}}\right) \tag{2.2}$$

$$\mathrm{PE}(\mathrm{pos}, 2i+1) = \cos\left(\frac{\mathrm{pos}}{10000^{\frac{2i}{d_{\mathrm{model}}}}}\right) \tag{2.3}$$

where pos is the position index, i is the dimension index, and d_{model} is the embedding dimension. These functions create a unique pattern for each position that the model can learn to interpret, enabling it to incorporate both semantic meaning and sequential position.

2.2.6 Processing Through Transformer Layers

The sequence of token embeddings is processed through multiple transformer layers, each refining the representation based on both local and global context. In each layer:

1. Self-attention mechanisms compute how each token should attend to allowed tokens
2. Multi-head attention allows the model to focus on different aspects of relationships
3. Feed-forward networks process each token's representation individually
4. Residual connections and normalization layers stabilize the processing.

This layered processing creates increasingly abstract and context aware representations, with deeper layers capturing more complex relationships and concepts. For technical applications in refining, this enables the model to connect surface level descriptions with deeper engineering principles and domain knowledge.

2.2.7 Self-attention: The Core Innovation

At the heart of the Transformer architecture is the self-attention mechanism, which allows the model to weigh the importance of different words in a sequence when computing a representation for each word. This mechanism can be understood as a sophisticated form of context awareness that determines how much focus to place on each word when processing any given word in the sequence.

The self-attention mechanism operates on three matrices derived from the input: Query (Q), Key (K), and Value (V). These matrices are created through linear transformations of the input embeddings and three matrices W^Q, W^K, and W^V, whose weights are learned during the training step. The attention scores are computed by taking the dot product of the Query and Key matrices, scaled by the square root of the dimension of the keys, and then passed through a softmax function to obtain attention weights, as probability like values between 0 and 1:

$$\text{Attention}(Q, K, V) = \text{softmax}\left(\frac{QK^T}{\sqrt{d_k}}\right)V \tag{2.4}$$

where

- Q represents what we're searching for
- K represents what we're comparing against
- V represents the information to retrieve
- d_k is the dimension of the keys, used for scaling

This seemingly abstract mathematical operation performs a remarkably intuitive function: it allows each word to "look at" other words in the sequence and gather information from them based on their relevance to the current word. This can be conceptualized as each word dynamically determining which other words provide the most valuable context for its interpretation.

2.2.8 Causal Masking in Decoder Architectures

A critical component in decoder-only architectures like GPT and Claude is causal masking (or autoregressive masking). While encoders allow each token to attend to all tokens in the sequence, decoders implement a causal mask that ensures each token can only attend to previous tokens and itself, but never to future tokens. This restriction is fundamental to autoregressive modeling, where the goal is to predict the next token based solely on previous tokens.

Mathematically, this is implemented by modifying the attention formula with a mask that sets attention scores for future positions to negative infinity before the softmax operation:

$$\text{Masked Attention}(Q, K, V) = \text{softmax}\left(\frac{QK^T}{\sqrt{d_k}} + M\right)V \tag{2.5}$$

where M is a mask matrix with:

$$M_{ij} = \begin{Bmatrix} (0 \text{ } \mathit{if} \text{ } i \geq j) \\ (-\infty \text{ } \mathit{if} \text{ } i < j) \end{Bmatrix} \tag{2.6}$$

This ensures that the attention weights for future positions become zero after the softmax, effectively preventing any information flow from future tokens to the current token. This causal masking is essential for the model's ability to generate coherent text one token at a time during both training and inference and is a defining characteristic of decoder-only architectures.

This mathematical operation performs a remarkably intuitive function: it allows each word to "look at" preceding words in the sequence and gather information from them based on their relevance to the current word. As in the "unmasked" case, this can be conceptualized as each word dynamically determining which other words provide the most valuable context for its interpretation.

Multi-head attention extends this concept by performing multiple attention operations in parallel. Each "head" can focus on different aspects of the relationships between words; some might capture syntactic dependencies, others semantic similarities, and others thematic relationships. This multi-faceted attention enables the model to capture a rich representation of the text.

2.2.9 Feed-Forward Networks in Transformer Architectures

Beyond the attention mechanism, each Transformer block contains a feed-forward neural network that processes each token's representation individually. This component transforms the attention-weighted information through a two-layer network with a non-linear activation function:

$$\text{FFN}(x) = \max(0, xW_1 + b_1)W_2 + b_2 \tag{2.7}$$

where

- x is the input vector from the previous sublayer
- W_1 and W_2 are learned weight matrices for the first and second linear transformations
- b_1 and b_2 are learned bias vectors
- $\max(0, \ldots)$ is the ReLU (Rectified Linear Unit) activation function

The first linear transformation $(xW_1 + b_1)$ projects the input into a higher-dimensional space, typically four times larger than the model dimension (e.g., if the embedding dimension is 768, this intermediate dimension might be around 3072). The ReLU function then introduces non-linearity by replacing negative values with zeros. Finally, the second linear transformation returns the representation to the original dimensionality.

This seemingly simple component is essential for the Transformer's capabilities, as it enables each position to process its information independently, complementing the contextual information obtained through attention. The feed-forward network effectively serves as a position wise fully connected layer, applying the same transformation to each position separately and identically.

Layer normalization stabilizes the learning process by normalizing the inputs across the feature dimension, similar to when process engineers normalize variables of different scales in a multivariate control system. This normalization helps maintain consistent statistical properties throughout the network, facilitating more stable and efficient training.

These architectural components, self-attention with causal masking, feed-forward networks, and normalization layers, are stacked multiple times (typically 12–96 layers in modern LLMs), creating a deep network capable of learning increasingly abstract representations of language.

2.2.10 Residual Connections and Normalization Layers

The Transformer architecture includes two crucial components that enable the training of deep networks: residual connections and normalization layers.

Residual Connections (Skip Connections)

Residual connections simply add the input of a sublayer to its output:

$$x_{\text{out}} = x_{\text{in}} + \text{Sublayer}(x_{\text{in}}) \tag{2.8}$$

These connections create shortcuts through which information can flow directly across layers. This design helps solve the "vanishing gradient" problem that traditionally made deep networks difficult to train. By providing direct paths for gradients to flow backward during training, residual connections ensure that even deep layers can receive meaningful updates.

Normalization Layers

Normalization layers standardize the input values within each layer, ensuring they maintain a similar scale throughout the network. In Transformers, layer normalization is typically used, which normalizes values across the feature dimension for each position:

$$\text{LayerNorm}(x) = \gamma * \frac{x - \mu}{\sigma} + \beta \tag{2.9}$$

where μ and σ are the mean and standard deviation, while γ and β are learnable parameters.

These two techniques complement each other perfectly:

- Residual connections allow information to flow freely through the network
- Normalization layers keep the numerical values stable.

Together, they allow transformers to be built with dozens or even hundreds of layers while remaining trainable. Without these components, deep transformer models would be unstable during training and fail to learn effectively.

This clever combination has become a standard practice in modern neural network design, enabling the development of increasingly deeper and more powerful language models.

2.2.11 Output Generation: The Autoregressive Process

In generative applications, which constitute most refining related use cases, the LLM produces outputs through an autoregressive process:

1. The model computes a probability distribution over its entire vocabulary for the next token
2. A token is selected based on this distribution, influenced by parameters like temperature (controlling randomness)
3. The selected token is appended to the input
4. The process repeats with this extended input to generate the next token.

This step-by-step generation process explains several key characteristics of LLM outputs. The probabilistic nature of token selection introduces some inherent variability in outputs, even with identical inputs. The autoregressive approach also means that early token selections influence all subsequent ones, creating a coherent flow but potentially propagating errors or biases.

2.2.12 Generation Parameters: Controlling Output Characteristics

Several parameters significantly influence the behavior of LLMs during generation, providing important controls for technical applications:

Temperature: Controls the randomness in token selection:

- Low temperature (0.1–0.5): More deterministic, conservative responses
- Medium temperature (0.6–0.8): Balanced creativity and precision
- High temperature (> 1.0): More diverse, creative, but potentially less accurate responses.

For refining applications requiring technical precision, temperatures in the 0.2–0.5 range typically offer the best balance of accuracy and useful variation.

Table 2.1 Recommended LLM parameter settings for refining applications

Application type	Temperature	Top-*p*	Tokens	Rationale
Process troubleshooting	0.2–0.3	0.85–0.9	1000–2000	Precision and factual accuracy are critical
Design ideation	0.6–0.7	0.92–0.97	2000–4000	Balance between creativity and feasibility
Procedure generation	0.1–0.2	0.80–0.85	1500–3000	Highest precision and consistency
Technical research synthesis	0.4–0.5	0.88–0.92	3000–6000	Balanced accuracy with comprehensive coverage
Risk assessment	0.3–0.4	0.85–0.9	2000–4000	Thorough exploration of possibilities with reliability

Source Author's recommendations based on experimental optimization

Top-*p* (Nucleus Sampling): Dynamically limits token selection to a subset whose cumulative probability exceeds threshold *p*:

- Higher values (0.9–1.0): Consider more possible tokens, including less probable ones
- Lower values (0.5–0.8): Focus on more probable tokens, reducing creativity but increasing predictability.

Max Tokens: Limits the length of generated responses, important for controlling scope and focusing on concise technical analysis.

These parameters can be strategically adjusted based on the specific requirements of different refining applications, as shown in Table 2.1.

User Control of Generation Parameters

Users can control these important generation parameters through various interfaces when interacting with LLMs:

API Interactions: When using LLMs programmatically via APIs (e.g., OpenAI API, Anthropic API), users can specify these parameters directly in their API calls.

Web Interfaces: When using web-based interfaces like ChatGPT, Claude, or similar platforms, these controls are typically available through:

- Advanced options menus (often indicated by a settings icon)
- "Creative" versus "Precise" mode toggles (which adjust these parameters behind the scenes)
- Custom instructions or system prompts that specify desired behavior.

For industrial users developing prompt engineering strategies, it's recommended to:

1. Experiment systematically with different parameter combinations for each application type
2. Document successful parameter settings for different task categories
3. Create standardized configurations for common technical use cases
4. Consider different parameter settings for exploratory versus production uses.

The appropriate configuration of these parameters can significantly enhance the value of LLMs for technical applications, balancing between creative exploration and precise, reliable outputs.

2.3 Training Large Language Models

The remarkable capabilities of LLMs emerge from their training methodology, which has evolved significantly from traditional supervised learning approaches. Understanding this process is important for appreciating both their capabilities and limitations in technical domains like refining (Radford et al., 2019).

2.3.1 Pretraining: Learning from Massive Text Corpora

LLMs are initially trained through self-supervised learning on enormous datasets of text from the internet, books, scientific papers, and other sources. Unlike supervised learning, which requires labelled data, this approach uses the inherent structure of language itself as supervision. Specifically, models are trained to predict the next word in a sequence given the preceding words, a task known as autoregressive modelling.

The objective function for this pretraining phase is to minimize the negative log-likelihood:

$$L = -\Sigma_i \log P(x_i | x <_i) \tag{2.10}$$

where $P(x_i|x_{<i})$ is the probability the model assigns to the correct token x_i given its preceding context $x < i$.

This conceptually simple task requires the model to develop a sophisticated understanding of language, including grammar, facts about the world, reasoning patterns, and domain-specific knowledge including technical concepts relevant to petroleum refining and chemical engineering. The scale of this pretraining is staggering, with models trained on trillions of tokens.

Table 2.2 illustrates the exponential growth in model size and training data, highlighting how quantitative scaling has led to qualitative improvements in capabilities.

Table 2.2 Evolution of large language model scale

Model	Release year	Parameters	Training data size	Notable capabilities for technical domains
BERT	2018	340 M	16 GB	Basic technical term understanding
GPT-2	2019	1.5 B	40 GB	Simple process descriptions
GPT-3	2020	175 B	570 GB	Basic problem-solving, code generation
PaLM	2022	540 B	780 GB	Enhanced reasoning, technical analysis
GPT-4	2023	Not disclosed	Not disclosed	Complex problem solving, multimodal reasoning
Claude 3 (Opus)	2024	Not disclosed	Not disclosed	Advanced reasoning, improved context understanding
GPT-4o	2024	Not disclosed	Not disclosed	Optimized performance, faster response time
Gemini 1.5 Pro	2024	Not disclosed	Not disclosed	Extended context window (up to 1 M tokens)

Source Compiled from published research papers and official company announcements

2.3.2 Fine-Tuning: Specialization for Specific Tasks

While pretraining on diverse data creates a model with broad capabilities, further refinement is often needed for optimal performance in specific domains or tasks. This is achieved through fine-tuning, where the pretrained model is further trained on more specialized datasets with specific objectives (Wei et al., 2022). Several fine-tuning approaches have proven particularly effective:

1. **Supervised Fine-tuning**: Training on human-labelled examples of desired outputs for specific inputs, such as technical Q&A pairs relevant to refining operations.
2. **Reinforcement Learning from Human Feedback (RLHF)**: Using human preferences to guide model behavior, where human evaluators rank different possible responses, and the model learns to generate higher ranked outputs.
3. **Instruction Tuning**: Training the model to follow explicit instructions in various formats, improving its ability to respond appropriately to different types of queries.

These fine-tuning approaches have been instrumental in adapting LLMs for technical domains, as shown by improved performance on specialized tasks through different stages of training and fine-tuning.

2.3.3 Training Infrastructure and Computational Requirements

The computational resources required for training modern LLMs are substantial. Training a model requires thousands of specialized GPUs (Graphics Processing Units) or TPUs (Tensor Processing Units) operating for months, with costs estimated in tens to hundreds of millions of dollars.

This extreme computational intensity has significant implications for the evolution of LLMs in specialized technical domains like refining. While generic models can be developed by organizations with massive computing resources, domain-specific fine-tuning offers a more accessible pathway for industry specific applications, leveraging the general capabilities of pretrained models while adapting them to specialized technical contexts.

2.4 Epistemological Limitations: Navigating the "Black Box" Challenge

In discussing the architecture and functioning of LLMs, it is essential to address a fundamental epistemological limitation that affects the entire discipline of prompt engineering: the inherent opacity of the causal mechanisms linking prompts to outputs. This limitation represents a significant departure from traditional engineering disciplines and requires explicit recognition.

2.4.1 The Challenge of Causal Understanding

While we understand the general architecture of transformer-based models, including attention mechanisms, positional encoding, and generation parameters, this understanding does not translate into deterministic predictions about how specific prompts will produce specific outputs. Several factors contribute to this limitation:

1. **Emergent Behavior from Complex Systems**: The behavior of LLMs emerges from interactions between billions or trillions of parameters, creating a system whose complexity exceeds our ability to trace causal pathways deterministically.
2. **Non-linearity and High Dimensionality**: The high dimensional, non-linear nature of neural networks makes it mathematically intractable to establish formal derivations linking inputs to outputs.

3. **Statistical rather than Deterministic Foundations**: LLMs operate on probabilistic principles, each token selection represents a sampling from a probability distribution rather than a deterministic function.
4. **Training Data Opacity**: The specific influences of different segments of the training data on model behavior remain largely unknown, even to the model's creators.

This opacity means that prompt engineering cannot currently claim the kind of mathematical rigor or causal predictability found in traditional engineering disciplines such as structural engineering or circuit design.

2.4.2 Implications for Prompt Engineering Methodology

Given these limitations, prompt engineering is more accurately characterized as an empirical, observational practice rather than a deductive science. Our approach in this book acknowledges this reality through several methodological commitments:

1. **Empirical Observation Over Theoretical Derivation**: We focus on documenting observed patterns of model behavior rather than attempting to derive them from first principles.
2. **Practical Utility Rather Than Theoretical Certainty**: The techniques presented are valued for their instrumental utility in specific contexts rather than as universal laws.
3. **Statistical Thinking**: We view prompt engineering outcomes in probabilistic terms, recognizing the inherent variance in results even with identical inputs.
4. **Falsifiability and Replication**: We prioritize transparent documentation of all parameters to enable replication of observed patterns.

This approach parallels other fields that have successfully navigated similar epistemological terrain, such as certain areas of medicine before mechanistic understanding, weather prediction, and complex systems science.

2.4.3 Research Frontiers in Interpretability

Current research in LLM interpretability offers some promising avenues for addressing these limitations:

1. **Mechanistic Interpretability**: Efforts to reverse engineer specific computational patterns within neural networks have shown limited but encouraging results for simple cases.

2. **Attention Pattern Analysis**: Studying attention patterns can reveal correlations between prompt structures and model focus, though without fully explaining the subsequent processing.
3. **Causal Mediation Analysis**: Experimental approaches that systematically vary components of prompts while measuring outputs can establish some limited causal insights.
4. **Activation Atlas Approaches**: Visualizing and analyzing activation patterns across network layers provide another window into model behavior.

However, these approaches remain in early stages and far from providing comprehensive causal accounts of LLM behavior.

2.4.4 Practical Approach to Rigorous Prompt Engineering

Given these constraints, our approach to rigor in prompt engineering emphasizes:

1. **Transparent Reporting**: Document all relevant parameters, including model version, temperature, top-*p* values, and context conditions.
2. **Pattern Recognition**: Identify recurring patterns in prompt-response relationships without necessarily claiming to understand their causal mechanisms.
3. **Contextual Reliability**: Emphasize the contexts in which specific techniques tend to produce reliable results while acknowledging their limitations.
4. **Continuous Empirical Validation**: We subject prompt engineering techniques to ongoing empirical testing rather than relying on theoretical derivations.

This empirically grounded approach enables practical progress despite the fundamental limitations in our causal understanding of LLM behavior.

2.4.5 Implications for Industrial Applications

For industrial applications like petroleum refining, petrochemicals, power generation, cement, fertilizers among others these epistemological limitations necessitate specific safeguards:

1. **Rigorous Verification Protocols**: All LLM outputs must undergo verification against established engineering principles and empirical data.
2. **Domain Expert Oversight**: Subject matter experts must play a central role in evaluating the validity and applicability of LLM generated content.

3. **Graduated Trust Framework**: Organizations should implement frameworks that assign appropriate levels of scrutiny based on the criticality and risk of different applications.
4. **Continuous Learning Systems**: Practical applications should incorporate feedback loops that track the reliability of different prompt engineering approaches in specific contexts.

By explicitly acknowledging these epistemological limitations and implementing appropriate safeguards, organizations can realize the benefits of prompt engineering while preserving the integrity of safety critical systems and decisions.

2.4.6 Future Research Directions

Moving toward greater causal understanding remains an important research goal. Future work may focus on:

1. **Controlled Experimentation**: Systematic studies that isolate specific prompt variables and measure their effects on outputs.
2. **Computational Models of Prompting**: Development of simplified models that might capture some aspects of prompt–response relationships.
3. **Transfer Learning Analysis**: Investigation of how prompt techniques transfer across different models and domains, potentially revealing underlying principles.
4. **Integration with Mechanistic Interpretability**: Combining prompt engineering with advances in neural network interpretability.

By pursuing these research directions while maintaining epistemological humility about our current understanding, the field can continue to advance in both practical utility and theoretical foundation.

2.5 Applications in Petroleum Refining and Petrochemical Processing

The capabilities of LLMs are valuable in a wide range of applications across the refining and petrochemical sectors. These applications span the entire value chain from research and development through design, operations, maintenance, and strategic planning.

2.5.1 Research and Development Applications

LLMs can significantly accelerate the R&D process in several ways:

1. **Literature Review and Synthesis**: Rapidly aggregating and summarizing relevant research across diverse sources, identifying key trends, contradictions, and knowledge gaps.
2. **Hypothesis Generation**: Suggesting novel research directions based on patterns and relationships that might not be immediately apparent to human researchers.
3. **Experimental Design**: Proposing experimental approaches and test matrices for evaluating new catalysts, process conditions, or feedstock combinations.
4. **Data Interpretation**: Assisting in the analysis of complex experimental results, identifying patterns and potential explanations.

These capabilities may compress R&D timelines by reducing the effort required for knowledge acquisition and experimental planning, allowing researchers to focus more on creative problem solving and critical analysis.

2.5.2 Process Design and Optimization

In process design applications, LLMs offer several advantages:

1. **Design Space Exploration**: Rapidly generating and evaluating multiple design concepts based on specified constraints and objectives.
2. **Process Integration**: Identifying opportunities for heat, mass, and work integration across process units that might be overlooked in traditional design approaches.
3. **Parameter Optimization**: Suggesting operating parameter combinations that balance multiple objectives like yield, energy efficiency, and product quality.
4. **Design Verification**: Identifying potential issues, conflicts, or oversights in proposed designs before detailed engineering.

Table 2.3 outlines example applications of LLMs across different design phases, from conceptual development through detailed engineering.

2.5.3 Operations Support and Troubleshooting

In operational contexts, LLMs can provide valuable support for both routine operations and exception handling:

Table 2.3 LLM applications across process design phases

Design phase	Key LLM applications	Value proposition
Conceptual design	Technology screening, process scheme generation, preliminary mass balance	Broader exploration of design space, identification of nonobvious alternatives
Basic engineering	Process flow development, equipment sizing guidelines, control philosophy development	Accelerated development of design basis, enhanced consistency across design elements
Detailed engineering	Specification development, design verification, constructability assessment	Error reduction, completeness verification, design optimization opportunities
Commissioning planning	Procedure development, risk assessment, Testing protocol development	More comprehensive planning, Identification of potential issues, knowledge transfer

Source Author's analysis based on industry applications

1. **Procedure Enhancement**: Developing, reviewing, and optimizing standard operating procedures for clarity, completeness, and safety.
2. **Troubleshooting Assistance**: Analyzing operational anomalies, suggesting potential causes, and recommending investigation approaches.
3. **Knowledge Management**: Capturing, organizing, and making accessible the knowledge of experienced operators and engineers.
4. **Decision Support**: Providing structured analysis of complex operational decisions, considering multiple factors and potential outcomes.

These applications are particularly valuable for addressing the knowledge transfer challenges facing the refining industry as experienced personnel retire, taking decades of operational expertise with them.

2.5.4 Strategic Planning

At the strategic level, LLMs can assist with the long-term challenges faced by the refining industry:

1. **Technology Scouting**: Analyzing emerging technologies, regulatory trends, and market shifts to develop comprehensive transformation strategies.
2. **Scenario Analysis**: Generating and evaluating multiple future scenarios to test the robustness of strategic plans.

3. **Sustainability Planning**: Developing pathways toward reduced carbon intensity, including process modifications, feedstock transitions, and carbon management approaches.
4. **Investment Prioritization**: Evaluating complex trade-offs in capital allocation across competing priorities like reliability, efficiency, growth, and sustainability.

2.5.5 Limitations and Responsible Implementation

While LLMs offer remarkable capabilities for refining applications, their effective implementation requires clear understanding of their limitations and careful attention to responsible use.

Technical Limitations
Several technical limitations affect the application of LLMs in refining contexts:

1. **Hallucinations**: When the information requested exceeds their training data, LLMs can generate apparently plausible but factually incorrect information, particularly for specific numerical values or detailed technical specifications. In refining applications, where decisions can have significant safety, environmental, and economic implications, this necessitates rigorous verification protocols.
2. **Reasoning Limitations**: While LLMs show impressive reasoning capabilities in many contexts, they struggle with complex mathematical calculations, physical simulations, and certain types of logical analysis that are routine in refining engineering.
3. **Temporal Knowledge Constraints**: The fixed training cutoff date means LLMs lack awareness of recent regulatory changes, technological advances, or market developments unless explicitly provided with updated information.
4. **Context Window Limitations**: Current models have finite context windows (typically 4000–100,000 tokens), limiting their ability to process very long documents or maintain context across extended discussions.
5. **Domain Adaptation Challenges**: General purpose LLMs may require significant fine-tuning or prompting optimization to perform optimally on specialized refining tasks.

These limitations suggest that AI-assisted design is best viewed as a complement to rather than replacement for conventional engineering expertise, particularly valuable for early concept development and knowledge integration rather than detailed design or implementation planning.

Implementation Framework for Refining Applications
Based on the capabilities and limitations of current LLMs, a structured implementation framework for refining applications includes:

1. **Task Suitability Assessment**: Carefully evaluating which tasks are appropriate for LLM assistance based on criticality, verification options, and complementary capabilities.
2. **Human-AI Collaboration Design**: Structuring workflows to leverage the complementary strengths of human experts and LLMs, maintaining clear responsibility boundaries.
3. **Verification Systems**: Implementing rigorous processes for validating LLM outputs, particularly for safety–critical applications or numerical specifications.
4. **Continuous Evaluation**: Regularly assessing both the performance of LLMs on specific tasks and their impact on overall work processes and outcomes.
5. **Organizational Knowledge Integration**: Developing methods to incorporate organizational knowledge and expertise not captured in LLM training data.

This implementation framework should be viewed as a continuous cycle of assessment, implementation, verification, and refinement.

Ethical and Organizational Considerations

Beyond technical implementation, several broader considerations deserve attention:

1. **Transparency**: Maintaining clear communication about when and how LLMs are being used, including their limitations and the verification processes in place.
2. **Expertise Development**: Ensuring that reliance on LLMs enhances rather organizational expertise, using them as tools for knowledge amplification rather than replacement.
3. **Bias Awareness**: Recognizing potential biases in LLM outputs, particularly for subjective judgments or areas with societal dimensions.
4. **Governance Framework**: Establishing clear governance structures for LLM implementation, including responsibility allocation, approval processes, and usage policies.
5. **Data Security**: Managing sensitive information in interactions with LLMs, particularly for proprietary processes or confidential information.

These considerations highlight that effective LLM implementation is not merely a technical challenge but also an organizational and cultural one, requiring thoughtful change management and clear communication.

2.6 Conclusion: The Path Forward for LLMs in Refining

Large Language Models represent a significant advance in artificial intelligence with relevance to the petroleum refining industry. Their ability to integrate knowledge across domains, aid in complex problems solving, and generate coherent technical content offers valuable capabilities for an industry facing substantial challenges.

The architecture and functionality of these models based on the Transformer's attention mechanisms and trained on vast amounts of data enable them to capture representations of language that include substantial technical knowledge relevant to refining operations. While they have important limitations, particularly around factual precision and mathematical reasoning, these can be managed through appropriate implementation frameworks and verification processes.

The most promising approach is neither uncritical enthusiasm nor excessive caution, but rather a balanced integration that leverages LLMs' strengths while compensating for their weaknesses. By combining LLM capabilities with human expertise, specialized engineering tools, and rigorous validation procedures, refining organizations can extract significant value while managing risks appropriately.

As LLM technology continues to evolve with improvements in reasoning capabilities, knowledge integration, and domain specialization, we can anticipate expanding applications throughout the refining value chain. The organizations that develop the capability to effectively leverage these tools, while maintaining strong domain expertise and critical judgment, will be best positioned to navigate the complex challenges of refineries in the decades ahead.

The epistemological limitations discussed in this chapter do not diminish the value or validity of prompt engineering, they merely situate it within a different paradigm of scientific and engineering rigor, work with LLMs may require new standards of rigor that explicitly acknowledge the emergent and partially opaque nature of these systems.

The approach presented in this book embraces this challenge by combining empirical observation, practical utility, transparent methodology, and continuous validation creating a form of rigor appropriate to the unique characteristics of large language models and their applications in industrial transformation.

References

Bommasani R, Hudson DA, Adeli E, Altman R, Arora S, von Arx S, Bernstein MS, Bohg J, Bosselut A, Brunskill E, Brynjolfsson E.et al. (2021) On the opportunities and risks of foundation models. arXiv preprint arXiv:2108.07258

Brown TB, Mann B, Ryder N, Subbiah M, Kaplan J, Dhariwal P, Neelakantan A, Shyam P, Sastry G, Askell A et al (2020) Language models are few-shot learners. Adv Neural Inf Process Syst 33:1877–1901

Chowdhery A, Narang S, Devlin J, Bosma M, Mishra G, Roberts A, Barham P, Chung HW, Sutton C, Gehrmann S, Schuh P (2022) PaLM: scaling language modeling with pathways. arXiv preprint arXiv:2204.02311

Clark K, Khandelwal U, Levy O, Manning CD (2019) What does BERT look at? An analysis of BERT's attention. arXiv preprint arXiv:1906.04341

Colback L (2024) AI and the R&D revolution. Financial Times. Published 27 Nov 2024

Radford A, Wu J, Child R, Luan D, Amodei D, Sutskever I (2019) Language models are unsupervised multitask learners. OpenAI Blog 1(8):9

Vaswani A, Shazeer N, Parmar N, Uszkoreit J, Jones L, Gomez AN, Kaiser Ł, Polosukhin I (2017) Attention is all you need. Adv Neural Inf Process Syst 30

Wei J, Bosma M, Zhao VY, Guu K, Yu AW, Lester B, Du N, Dai AM, Le QV (2022) Finetuned language models are zero-shot learners. arXiv preprint arXiv:2109.01652

3 Technological Advances in Refining: Evolution and Innovation Toward Net Zero Refineries

Abstract

The petroleum refining industry has undergone continuous technological evolution since the mid-nineteenth century, driven by changing market dynamics, environmental regulations, energy crises, and societal demands. This chapter examines the technological progression from simple distillation through increasingly sophisticated conversion processes, identifying patterns of innovation that provide valuable insights for contemporary net zero transformation. The analysis traces the development of key refining technologies including atmospheric and vacuum distillation, thermal and catalytic cracking, reforming, hydroprocessing, and energy integration systems. Each technological advancement emerged in response to specific challenges while being enabled by concurrent scientific developments and economic imperatives. The chapter shows how external pressures have consistently impulsed innovation acceleration, with crisis periods driving revolutionary improvements in energy efficiency and environmental performance. Modern developments in biorefining, digitalization, and carbon management are examined as precursors to net zero refining systems. The analysis reveals that successful technology transitions have typically occurred through phased implementation rather than revolutionary replacement, providing a model for contemporary sustainability transformation. The chapter concludes by identifying emerging technologies critical for net zero refineries, including advanced

Rafael Larraz, Prompt Engineering and the Transformation of Petroleum Refining.

R. Larraz, *Prompt Engineering and the Transformation of Petroleum Refining*, Synthesis Lectures on Chemical Engineering and Biochemical Engineering,
https://doi.org/10.1007/978-3-031-99728-0_3

automated sorting, catalytic pyrolysis, biomass gasification, Fischer–Tropsch synthesis, and CO_2 utilization systems. These historical innovation patterns suggest that the transition to net zero refining, while challenging, follows established trajectories of progressive transformation driven by external imperatives and enabled by technological convergence.

3.1 Introduction

The petroleum refining industry has experienced a profound technological evolution since its inception in the mid-nineteenth century. This transformation represents one of the most significant examples of continuous industrial innovation, driven by changing market dynamics, environmental regulations, energy crises, and societal demands, as shown in Fig. 3.1, which illustrates the convergence of enabling factors for refining innovation. Each technological advancement has emerged in response to specific challenges, enabled by concurrent scientific developments and shaped by economic imperatives (Yergin 2011).

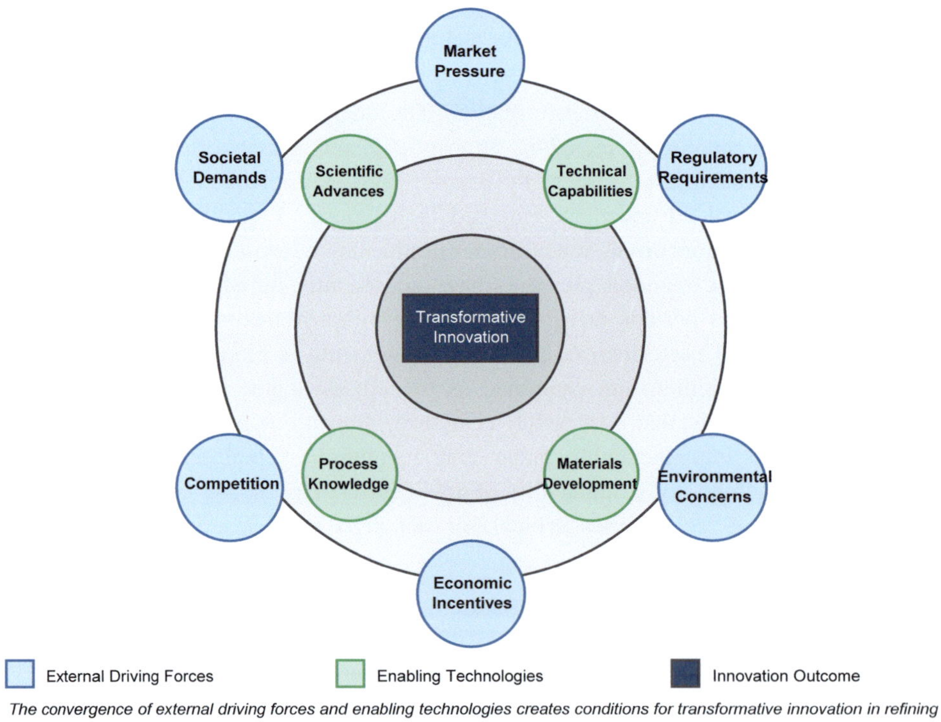

Fig. 3.1 Convergence of enabling factors for refining innovation

As the global energy system confronts the decarbonization mandate, the refining industry faces perhaps its most significant challenge ever: the transition to net zero emissions operations. This transformation must occur while maintaining economic viability and continuing to meet society's ongoing demand for liquid fuels and petrochemical products. Understanding the patterns of historical innovation in refining can provide valuable insights for navigating this complex transition (Speight 2011).

This chapter examines the technological evolution of petroleum refining through a structured analysis of major innovations, identifying the contextual factors that impulse each advancement and extracting principles relevant to net zero refineries. Rather than simply cataloguing technologies, this analysis seeks to uncover the underlying innovation mechanisms that have repeatedly enabled the industry to overcome past challenges (Jones and Pujadó 2006).

After an overview of the refining industry, the analysis is organized chronologically and thematically, beginning with the fundamental separation processes that defined early refining and progressing through increasingly sophisticated conversion, treatment, and optimization technologies. The chapter concludes with an examination of emerging net zero technologies that represent the next frontier of refining innovation.

3.2 The Petroleum Refining Industry: An Overview

Petroleum refining represents one of the world's most critical industrial processes, serving as the bridge between crude oil extraction and the set of products that power modern civilization. With approximately 700 refineries operating globally and a combined processing capacity exceeding 100 million barrels per day, this industry forms the backbone of global energy infrastructure and chemical manufacturing. The United States, China, Russia, India, and Japan collectively host about 40% of global refining capacity, reflecting both historical development patterns and contemporary energy demands.

Modern refineries vary significantly in size, from small operations processing less than 50,000 barrels per day to massive integrated complexes handling over 1 million barrels daily. These facilities range from simple hydroskimming refineries performing basic separation and treating processes to highly complex conversion refineries capable of transforming the heaviest crude oil fractions into high-value transportation fuels and petrochemical feedstocks.

The economic significance of refining extends far beyond the direct value addition of converting crude oil into products. As a capital-intensive industry with typical facility investments ranging from $1 to $10 billion, refineries create substantial employment, support auxiliary industries, and generate significant tax revenue. Moreover, refining economics profoundly influence global energy markets, with refinery margins (the difference between crude oil costs and product values) serving as critical indicators of energy market health.

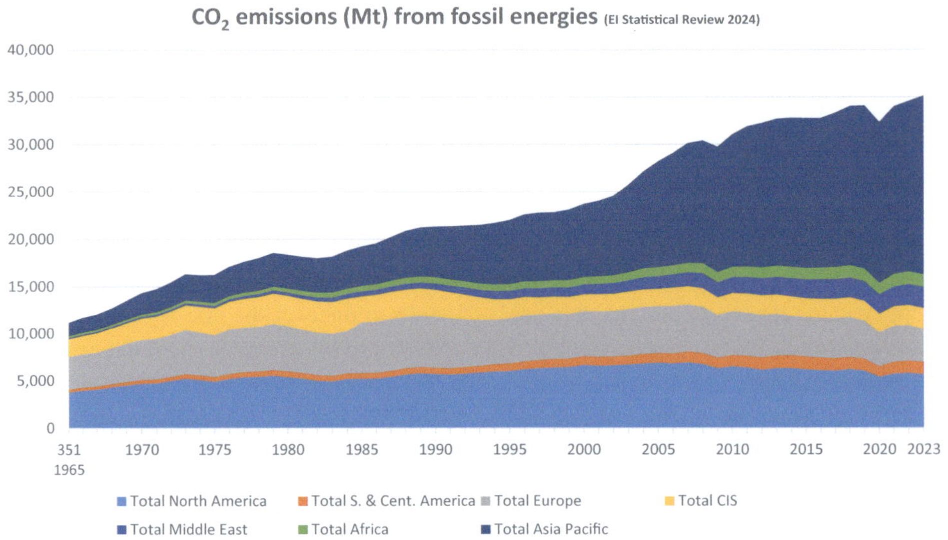

Fig. 3.2 CO_2 emissions evolution from fossil energies

Figure 3.2 shows the evolution of CO_2 emissions sourced by fossil energies. The petroleum refining industry represents a significant contributor to global CO_2 emissions, accounting for approximately 31% or 11,800 million tons per year of all industrial carbon dioxide releases worldwide. As a highly energy-intensive sector, refineries consume vast amounts of fossil fuels to power the complex processes that transform crude oil into usable products like gasoline, diesel, and petrochemicals. Despite technological improvements in efficiency, the industry still faces substantial challenges in reducing its carbon footprint. The continued growth in global demand for refined petroleum products, particularly in developing economies, has offset many of the efficiency gains achieved in recent decades.

3.2.1 Basic Refinery Configuration and Operation

At its core, a petroleum refinery separates and transforms crude oil, a complex mixture of hundreds of thousands of hydrocarbon compounds, into valuable products through a series of physical and chemical processes. While each refinery has a unique configuration based on factors including crude feedstock properties, desired product slate, market demands, and regulatory requirements, most modern facilities share common fundamental process units arranged in a sequential flow.

The refining process typically begins with crude oil desalting and atmospheric distillation, where incoming crude is heated and separated into various fractions based on boiling ranges. These initial fractions, where C stands for typical number of carbon atoms of fraction's molecules, include:

- Light gases (C_1-C_4): Used as refinery fuel or processed into liquefied petroleum gas (LPG)
- Naphtha (C_5-C_{12}): Precursor to gasoline and petrochemical feedstock
- Kerosene (C_{10}-C_{16}): Processed into jet fuel and heating oil
- Diesel/Gas-oil (C_{14}-C_{20}): Refined into diesel fuel and heating oils
- Residue (> C_{20}): Heavy material further processed or sold as fuel oil.

The atmospheric distillation residue typically undergoes vacuum distillation, which separates additional gas oil fractions under reduced pressure to avoid thermal decomposition. From these primary separation processes, as shown in Fig. 3.3, various intermediate streams flow to conversion, upgrading, and treating units that modify molecular structures and remove impurities, typically by means of catalysts, which have been the driving force of technological development.

Key secondary processing units in modern refineries include:

- **Catalytic reforming**: Converts low-octane naphtha into high-octane gasoline components
- **Fluid catalytic cracking (FCC)**: Transforms heavy gas oils into gasoline, distillates, and lighter products
- **Hydrocracking**: Combines hydrogen with heavy fractions to produce lighter products
- **Hydrotreating**: Removes sulfur, nitrogen, and other contaminants using hydrogen

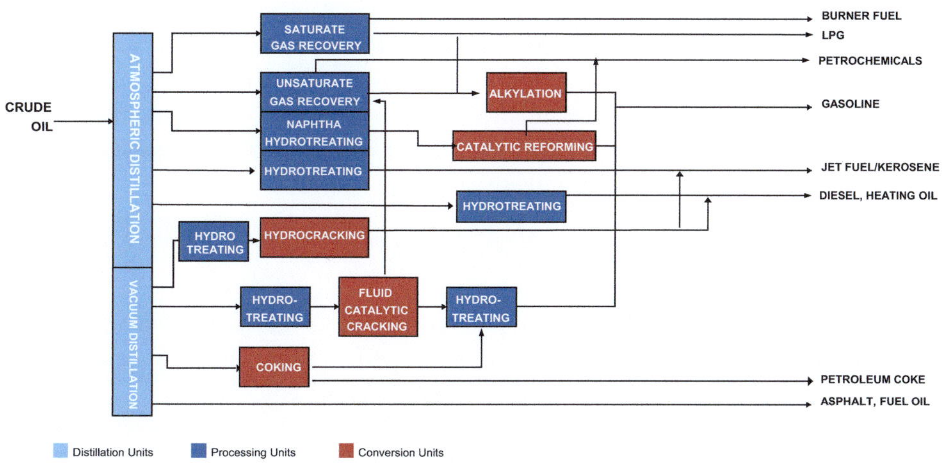

Fig. 3.3 Simplified flowchart of a complex refinery

- **Alkylation**: Combines light olefins with isobutane to produce high-octane gasoline components
- **Isomerization**: Rearranges straight-chain molecules into branched structures with improved properties
- **Coking**: Thermally cracks the heaviest residual fractions into lighter products and petroleum coke.

The products from these processes are then blended into final products meeting precise specifications for market distribution. A modern complex refinery simultaneously produces multiple product streams, including:

- Transportation fuels (gasoline, diesel, jet fuel): Typically 70–80% of output
- Heating fuels (heating oil, fuel oil, LPG): 10–15% of output
- Petrochemical feedstocks (naphtha, aromatics): 5–10% of output
- Specialty products (lubricants, waxes, asphalt): 5–10% of output.

Supporting these core processing units, refineries exhibit extensive utility systems providing steam, hydrogen, power, cooling water, and compressed air, along with auxiliary facilities for wastewater treatment, sulfur recovery, and product storage and handling. Modern refineries also incorporate advanced control systems, enabling real-time optimization of operations to maximize efficiency and product value.

3.2.2 Industry Challenges and Evolution Drivers

The refining industry has continuously evolved in response to changing market demands, regulatory requirements, and technological capabilities. Several fundamental drivers, as previously shown in Fig. 3.1, have shaped its development trajectory:

1. **Market demand shifts**: The progressive increase in demand for transportation fuels, especially gasoline and middle distillates (diesel and jet fuel), has pushed refineries toward configurations maximizing these products.
2. **Feedstock quality deterioration**: The global crude slate has generally become heavier (higher density) and sourer (higher sulfur content) over time, requiring more deeper conversion and treating technologies.
3. **Product specification tightening**: Increasingly stringent regulations on fuel properties, particularly sulfur content, aromatic levels, and other environmental parameters have necessitated substantial investments in advanced treating processes.
4. **Environmental regulations**: Progressively tighter controls on emissions, effluents, and waste products have driven innovation in both process technologies and end-of-pipe treatment systems.

5. **Energy efficiency imperatives**: Rising energy costs and sustainability concerns have motivated continuous improvements in thermal efficiency, heat integration, and overall energy management.
6. **Competition and margin pressure**: Chronic overcapacity in many markets has created intense competition, incentivizing operational excellence and strategic differentiation.

These drivers have collectively pushed the industry toward greater complexity, higher conversion capabilities, improved efficiency, and reduced environmental impact, a journey that continues today as the sector confronts decarbonization in response to climate change. This chapter examines the technological evolution of petroleum refining through a structured analysis of major innovations, identifying the contextual factors that drive each advancement and extracting principles relevant to net zero refineries. Rather than cataloguing technologies this analysis, shown in Fig. 3.4, seeks to uncover the underlying innovation mechanisms that have repeatedly enabled the industry to overcome past challenges (Jones and Pujadó 2006).

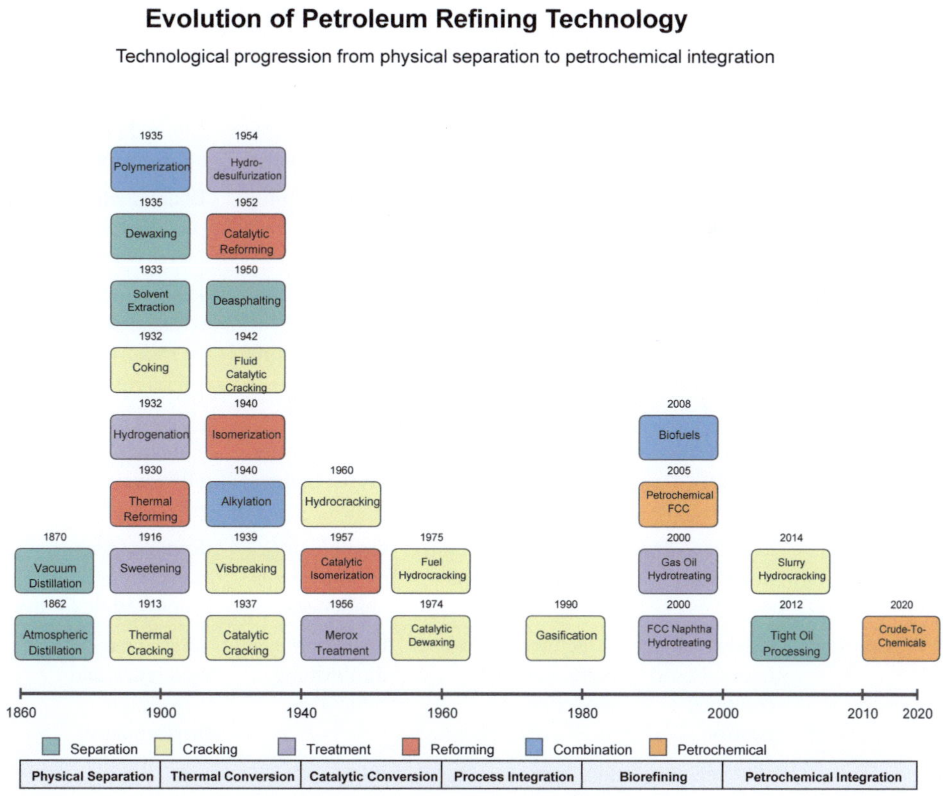

Fig. 3.4 Evolution of petroleum refining technology: 1860–2020

3.3 Atmospheric and Vacuum Distillation

Distillation represents the foundational technology of petroleum refining, establishing the basic principle of physical separation that would guide subsequent developments. Prior to systematic petroleum distillation, the procurement of petroleum derived products was rudimentary, focused primarily on kerosene for illumination. Early methods involved simple heating in open or semi-enclosed vessels, with limited fraction separation and considerable resource wastage (Forbes 1958). Table 3.1 summarizes the evolution of petroleum distillation technology from primitive batch stills to modern advanced systems with computer control.

The development of systematic distillation was prompted by several concurrent factors. Growing demand for illumination alternatives to increasingly scarce whale oil created economic incentives for kerosene production. The discovery of commercially viable oil fields, beginning with Edwin Drake's well in Pennsylvania in 1859, provided consistent raw material supply. The transfer of distillation knowledge from other industries like alcohol and essential oil production provided important technical foundations. Economic

Table 3.1 Evolution of petroleum distillation technology

Period	Technology	Key features	Limitations	Impact
Pre-1860s	Primitive batch stills	Simple heating, minimal fractionation	Low yields, poor quality control	Limited primarily to kerosene production
1860s-1910s	Continuous atmospheric stills	Continuous operation, improved fractionation	Limited ability to process heavy fractions	Enabled consistent product quality and increased throughput
1910s-1930s	Vacuum distillation	Enabled recovery of lubricating oils and heavier fractions at lower temperatures	Energy intensive, limited thermal cracking efficiency	Processing of reduced crude
1930s-1950s	Improved column designs	Better tray designs, reflux control	High pressure drops, fouling issues	Enhanced separation efficiency and product yield
1950s-1970s	Enhanced heat integration	Multiple feed points, stripping sections	Complexity, control challenges	Reduced energy consumption and improved flexibility
1970s-present	Advanced internals and computer control	Structured packings, computational modeling	Capital intensity, diminishing returns	Optimized energy efficiency and separation performance

Source Forbes (1958), Speight (2011), Jones and Pujadó (2006)

pressures to maximize resource utilization drove continuous improvements in separation efficiency (Chandler 1990).

The separation of crude oil heating and separation stages after the innovative process introduced by Milo Trumble in 1910 and the implementation of vacuum distillation in the early 1920s represented a significant advancement. By reducing operating pressure, refiners could distill heavier fractions at lower temperatures, avoiding thermal cracking and enabling the recovery of valuable lubricating oils. This development established the principle that modifying process conditions could overcome apparent limitations, a concept with continuing relevance for low-carbon process design (Nelson 1958).

Modern distillation units have evolved substantially from their early predecessors, incorporating enhanced heat integration, advanced internals, and control systems. The development of structured packings significantly improved separation efficiency while reducing pressure drop. Computer modeling enabled optimization of complex distillation trains, maximizing resource efficiency. These improvements collectively reduced the energy intensity of distillation by approximately 30–50% compared to early twentieth century designs (Jones and Pujadó 2006).

The evolution of distillation technology offers several insights relevant to net zero refinery development. The principle of comprehensive resource utilization maximizing value from each fraction of crude oil parallels the need to integrate and valorize all streams in a net zero facility, including carbon dioxide emissions. Additionally, the transfer of technologies from adjacent industries accelerated distillation development, highlighting the potential value of cross-sectoral knowledge exchange for decarbonization.

3.4 Thermal Cracking

Thermal cracking represented the first significant departure from purely physical separation processes in refining, introducing chemical conversion to modify molecular structures and product yields. Prior to thermal cracking, refineries were constrained to the products directly obtainable through distillation, with relatively fixed gasoline yields.

The development of thermal cracking was driven by several interconnected factors. The automotive industry created substantial demand for gasoline, exceeding what simple distillation could provide. This market imbalance generated strong economic incentives for increased gasoline production. Concurrent scientific advances improved understanding of hydrocarbon chemistry and the effects of heat and pressure on molecular structures. Industrial competition, particularly among major oil companies, accelerated development as firms sought competitive advantages (Enos 1962).

The Burton process, developed by William Burton at Standard Oil of Indiana in 1913, represented the first commercially successful thermal cracking technology. This batch process subjected heavy oils to temperatures around 400 °C and pressures of 600–700 psi, nearly doubling gasoline yields compared to straight run distillation. The subsequent

Dubbs process, introduced in 1922, improved upon this foundation with continuous operation and liquid recycle streams, significantly enhancing operational efficiency and reducing downtime (Enos 1962).

Table 3.2 documents the evolution of thermal cracking processes from the initial Burton process through modern fluid coking systems.

Later developments in thermal cracking diversified to address specific objectives. Visbreaking emerged as a lower severity process focused on reducing the viscosity of heavy residues for their use as marine fuel. Delayed coking, meanwhile, accepted coke formation as inevitable and designed around it, creating a controllable process for converting residuals to lighter products plus petroleum coke. Fluid coking, introduced in the 1950s, represented the culmination of thermal cracking development, with continuous coke removal and improved heat transfer through fluidized bed technology (Speight 2011).

Table 3.2 Evolution of thermal cracking processes

Process	Developer	Year	Operating conditions	Key innovations	Limitations
Burton process	Standard oil (Indiana)	1913	Batch, 400 °C 70–100 psi	First commercial thermal cracking	Batch operation, limited conversion, frequent cleaning
Dubbs process	Universal oil products	1922	Continuous, 450–460 °C 70–100 psi	Continuous operation, liquid-phase recycle	Coking, limited selectivity
Cross process	Cross engineering	1929	470–500 °C 70–100 psi	Vapor-phase reaction, improved heat transfer	High energy consumption
Visbreaking	Various	1930s	450–500 °C, 50–200 psi	Lower severity for viscosity reduction	Limited conversion, product stability issues
Delayed coking	Standard oil (Indiana)	1930s	480–500 °C, 30–90 psi	Controlled coke formation, increased liquid yields	Solid coke handling, high energy intensity
Fluid coking	Exxon	1950s	480–565 °C, 10–50 psi	Continuous coke removal, improved heat transfer	Complex mechanical systems, emissions

Source Enos (1962), Gary et al. (2007), Jones and Pujadó (2006)

The thermal cracking era offers several instructive lessons for net zero refinery development. The innovation trajectory was driven by market imbalanced demand for a product (gasoline) exceeded what existing technology could supply. The contemporary parallel is clear: demand for low-carbon fuels and products now exceeds what conventional petroleum technology can deliver, creating similar innovation pressures.

The valorization of byproducts also emerged as a critical economic factor. As thermal cracking evolved, petroleum coke transitioned from problematic waste to valuable product for carbon anodes, fuel, and later, specialty graphites (Speight 2011). This principle applies directly to carbon capture and utilization strategies, where economic viability often depends on finding valuable applications for captured carbon dioxide.

3.5 Catalytic Cracking

Catalytic cracking represents one of the most important innovations in refining history, fundamentally altering the industry's capabilities and economics. Before catalytic cracking, thermal processes presented significant limitations: high energy consumption, limited control over product distribution, and moderate gasoline quality. The selectivity of thermal cracking was inherently constrained by reaction thermodynamics, generating substantial low value byproducts (Sadeghbeigi 2012).

The development of catalytic cracking emerged from a confluence of scientific, economic, and geopolitical factors. World War II created substantial demand for high octane aviation fuels, making improved conversion processes a strategic priority. Academic and industrial research advanced the understanding of acid catalysis and surface reaction mechanisms. The limitations of thermal cracking, particularly in product quality and selectivity, created economic incentives for improved processes. Collaborative efforts between petroleum companies, catalyst manufacturers, and research institutions accelerated development during wartime (Meyers 2004).

The initial fixed bed Houdry process, commercialized in 1937, demonstrated the commercial viability of catalytic cracking but faced operational limitations due to catalyst regeneration requirements. The fluid catalytic cracking (FCC) process, developed through the collaborative Catalytic Research Associates program and first commercialized in 1942, revolutionized the industry with continuous catalyst circulation between reaction and regeneration zones (Sadeghbeigi 2012). The catalyst fluidization and its continuous reaction-regeneration cycles, as shown in Fig. 3.5, is one of the most innovative and clever developments in the refining industry.

Subsequent developments in FCC technology focused on catalyst improvements, reactor-regenerator configurations, and process optimization. The introduction of synthetic silica-alumina catalysts in the 1950s improved activity and stability compared to natural clay catalysts. The significant advance introduction of zeolite catalysts in the

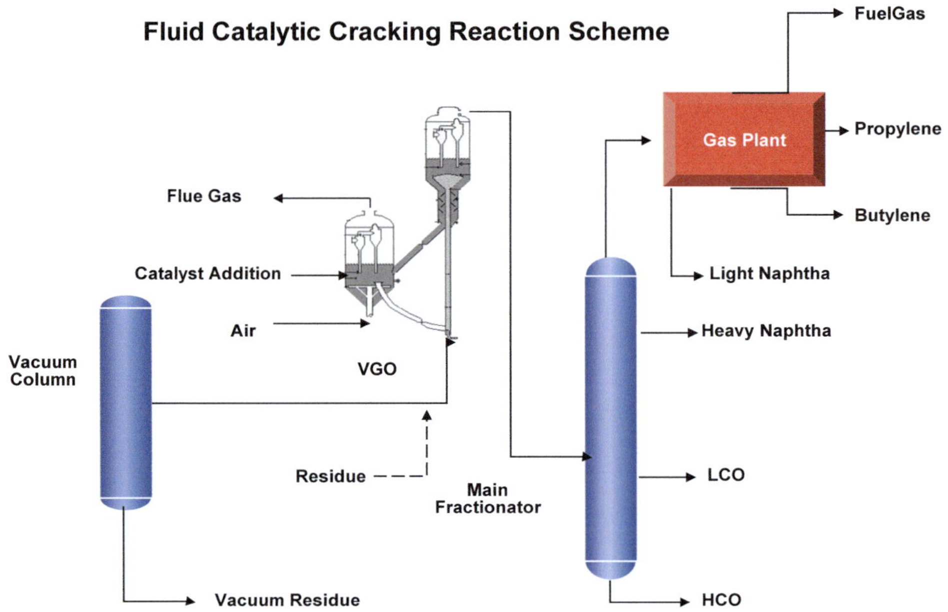

Fig. 3.5 Fluid catalytic cracking process scheme

1960s dramatically enhanced activity, selectivity, and gasoline yield, improving the process economics. Recent decades have seen further refinements in catalyst formulation, reactor design, and regenerator technology to address changing feedstock characteristics and product requirements (Occelli 2010). Table 3.3 documents the evolution of catalytic cracking catalysts from natural clays to modern environmental catalysts.

The commercial success of fluid catalytic cracking created abundant streams of light olefins, particularly propylene, isobutene and isoamylenes, that became valuable feedstocks for downstream molecular transformation processes. These olefin-rich streams, initially viewed as byproducts, evolved into strategic assets driving development of alkylation, oligomerization, and etherification technologies that maximized gasoline pool octane while addressing emerging environmental requirements.

Alkylation processes, developed in parallel with FCC technology, converted isobutane and FCC light olefins into high-octane branched paraffins essential for aviation gasoline during World War II and subsequently for automotive gasoline. The availability of consistent butene supplies from FCC units enabled large-scale alkylation implementation, creating synergistic value from what might otherwise have been low valued gas streams.

Oligomerization technology emerged to convert FCC propylene and butylenes into gasoline-range hydrocarbons through controlled polymerization reactions. Catalytic oligomerization processes, using solid phosphoric acid or zeolite catalysts, enabled conversion of C_3-C_5 olefins into C_6-C_{12} gasoline components with acceptable octane

Table 3.3 Evolution of catalytic cracking catalysts

Catalyst type	Period	Composition	Advantages	Limitations	Impact on process
Natural clays	1930s	Activated bentonite, montmorillonite	Available, inexpensive	Low activity, rapid deactivation	Required high catalyst-to-oil ratios
Synthetic amorphous	1940s–1950s	Silica-alumina, silica-magnesia	Improved selectivity, stability	Limited activity and sensitive to metals	Enabled practical FCC operation
Zeolite-containing	1960s–1970s	Y-zeolite in silica-alumina matrix	High activity, improved selectivity	Higher cost, sensitivity to poisons	Doubled gasoline yield, lower temperatures
Rare earth zeolites	1970s–1980s	RE-exchanged Y-zeolite	Enhanced stability, gasoline octane	Increased coke make, higher regeneration temperatures	Improved catalyst life in presence of metals
Ultra-stable zeolites	1980s-present	Dealuminated Y-zeolite, ZSM-5 additives	Metals tolerance, tailored selectivity	Complex manufacturing, higher cost	Flexibility for different feedstocks and product slates
Environmental catalysts	1990s–present	SOx/NOx reduction additives, low-metals transfer	Reduced emissions, longer cycles	Activity dilution, system complexity	Enabled compliance with environmental regulations

Source Occelli (2010), Sadeghbeigi (2012), Vogt and Weckhuysen (2015)

characteristics. The technology demonstrated how molecular assembly from small building blocks could create valuable products, establishing precedents for constructive use of light hydrocarbon streams rather than their disposal as fuel gas.

Etherification processes transformed FCC derived isobutene and isoamylenes into high-octane oxygenates as environmental regulations mandated cleaner-burning fuels. MTBE and TAME production from FCC olefins and methanol created valuable gasoline components while addressing air quality requirements. The development of reactive distillation for etherification, combining reaction and separation in single units, demonstrated process intensification principles directly applicable to integrated carbon capture and utilization systems. The subsequent evolution toward ETBE using ethanol instead of methanol illustrated technological adaptability in response to sustainability concerns.

The catalytic cracking revolution offers several crucial insights for net zero refinery development. Most fundamentally, it demonstrates the power of catalysis as an efficiency multiplier. Catalysts enabled processes that were more selective, operated at lower temperatures, and produced higher-value products compared to thermal routes. This principle is directly applicable to carbon management, where catalytic approaches to CO_2 conversion could similarly improve efficiency and economics (Vogt and Weckhuysen 2015).

The development history of FCC also highlights the value of collaborative innovation models. The wartime consortium that developed fluid catalytic cracking pooled intellectual and financial resources across companies that were otherwise competitors. Similar pre-competitive collaboration may be essential for addressing the fundamental challenges of refinery decarbonization, particularly for technologies with broad application but uncertain individual returns.

3.6 Catalytic Reforming

Catalytic reforming emerged as a critical technology for improving gasoline quality and providing petrochemical feedstocks. Before its development, refineries struggled to produce high octane gasoline without lead additives. Straight run and cracked naphtha had limited antiknock properties for increasingly demanding high compression engines. Simultaneously, growing petrochemical industries required reliable aromatic feedstocks (benzene, toluene, xylenes) that were difficult to produce efficiently (Antos and Aitani 2004). Catalytic reforming produces hydrogen as a by-product, this fact launched the development of several chemical processes linked to the cheap hydrogen availability as hydrotreating of petroleum streams to remove contaminants like sulfur, nitrogen and metals.

Commercial catalytic reforming processes emerged in the late 1940s and early 1950s. The pioneering Platforming process, developed by Universal Oil Products, utilized a platinum catalyst on alumina support, in a very innovative application of bifunctional catalysis. Competitive processes including Powerforming (Exxon), Ultraforming (Amoco), and Rheniforming (Chevron) followed with variations in reactor configuration, catalyst formulation, and operating conditions (Antos and Aitani 2004).

A significant breakthrough occurred in 1971 with UOP's introduction of Continuous Catalyst Regeneration (CCR) reforming. This technology-enabled continuous operation without the periodic shutdowns required by earlier fixed-bed processes. The constant removal, regeneration, and return of catalyst maintained higher average activity and selectivity while eliminating production interruptions. The CCR concept has since been adapted by most major technology licensors and represents the industry standard for new installations (Antos and Aitani 2004).

Catalyst development has remained a primary focus of reforming innovation. The introduction of rhenium as a catalyst promoter in the 1960s significantly improved catalyst stability. Subsequent development of multimetallic formulations incorporating elements such as tin, germanium, and iridium further enhanced performance for specific applications. Modern catalysts balance multiple functions: the acidic alumina support facilitates isomerization and cracking reactions, while the metal components catalyze dehydrogenation and hydrogenation steps (Antos and Aitani, 2004). Table 3.4 presents the major commercial catalytic reforming processes developed from 1949 to the 1990s.

Table 3.4 Major commercial catalytic reforming processes

Process	Developer	Year	Catalyst	Key features	Operating mode
Platforming	UOP	1949	Pt/Al_2O_3	First commercial process	Fixed-bed, cyclic
Powerforming	Exxon	1950	Pt/Al_2O_3	Improved regeneration system	Fixed-bed, cyclic
Ultraforming	Standard oil (Indiana)	1953	Pt/Al_2O_3, promoters	Higher severity operation	Fixed-bed, cyclic
Rheniforming	Chevron	1955	$Pt\text{-}Re/Al_2O_3$	Improved stability, selectivity	Fixed-bed, cyclic
Magnaforming	Engelhard	1968	Pt-multimetal	Enhanced metals resistance	Fixed-bed, cyclic
CCR platforming	UOP	1971	$Pt\text{-}Re/Al_2O_3$	Continuous catalyst regeneration	Continuous moving bed
Octanizing	IFP	1973	Pt-multimetal	Low-pressure operation	Fixed-bed, cyclic
AROMAX	CPC	1990s	Zeolitic	Enhanced aromatics production	Fixed-bed, cyclic

Source Antos and Aitani (2004), Meyers (2004)

The development of catalytic reforming created abundant supplies of hydrogen that enabled parallel advancement of isomerization technology for light naphtha streams. While reforming enhanced octane ratings of C_6-C_8 naphtha through aromatization and dehydrogenation, the light C_4-C_6 paraffins required different treatment to achieve acceptable gasoline blending characteristics. Isomerization processes, utilizing the hydrogen produced by reforming units, converted straight-chain paraffins into their branched isomers with a higher octane number, without changing carbon number or requiring additional hydrogen consumption.

The synergy between reforming and isomerization exemplified integrated refinery design principles that maximized value from each hydrocarbon stream. Reforming units provided both the hydrogen atmosphere and the economic incentive for isomerization development, as refiners sought to optimize their entire naphtha pool rather than individual fractions. Early isomerization processes used chlorided platinum catalysts similar to reforming, benefiting from shared catalyst expertise and operational experience. The evolution toward more stable zeolitic and sulfated zirconia catalysts paralleled reforming catalyst development, demonstrating how advances in one technology often enabled improvements in complementary processes.

This integration of reforming and isomerization established important precedents for molecular transformation without elemental composition changes reorganizing atoms within molecules rather than breaking bonds or introducing foreign elements.

The catalytic reforming development offers several valuable lessons for net zero refinery development. Perhaps most significantly, reforming demonstrates the principle of process versatility simultaneously addressing multiple objectives (octane improvement, hydrogen and aromatics production) through a single technology. This principle is directly applicable to decarbonization, where technologies that simultaneously reduce emissions and improve product properties will likely deliver superior economics (Antos and Aitani 2004). Also, the use of a precious metal like platinum in commercial catalysis opened a new way that can be extrapolated to the use of iridium in electrolyzers for green hydrogen production.

The catalyst regeneration evolution from cyclical regeneration to continuous systems illustrates how operational continuity can potentially enhance process economics. The challenge of integrating intermittent renewable energy into continuous refining operations presents a similar discontinuity that innovations in energy storage, hydrogen buffering, and flexible operations must address.

The upgrading of a low value stream in the isomerization porcess, has direct relevance to carbon management strategies in net zero refineries, where transformation rather than elimination of carbon containing molecules becomes key. The hydrogen-rich environment created by reforming operations provided the reducing atmosphere necessary for stable isomerization catalyst operation, illustrating how byproducts from one process can enable entirely different chemical transformations.

3.7 Hydroprocessing

The development of hydroprocessing represented a pivotal advancement in petroleum refining technology. Prior to hydrocracking and hydrotreating technologies, refineries faced significant limitations in converting heavy fractions and removing impurities such as sulfur, nitrogen, and metals. Refined products contained elevated levels of contaminants, and their quality was inconsistent and often suboptimal for increasingly sophisticated applications (Robinson and Dolbear 2006).

The technological landscape began to shift in the 1950s with the emergence of the first commercial hydrotreating units. This advancement was promoted by several concurrent factors. First of all, the catalytic reforming of naphtha developed a little bit earlier produces huge amounts of hydrogen ready to use. Early environmental regulations began imposing restrictions on sulfur content in fuels, particularly in urban areas where air quality concerns were growing. Simultaneously, the global crude slate was deteriorating in quality, with refiners processing increasingly heavier crudes with higher sulfur and contaminant levels (Scherzer and Gruia 1996). These challenges were compounded by advancements in metallurgical science, which enabled the development of materials capable of withstanding the demanding conditions of hydrogen rich environments at elevated temperatures and pressures.

The progression of hydroprocessing technology occurred in distinct phases. Initial hydrotreating processes focused on simple desulfurization of distillate streams, utilizing cobalt-molybdenum catalysts supported on alumina. These catalysts facilitated the selective hydrogenation of sulfur compounds, converting them to hydrogen sulfide which could be subsequently removed by means of ethanoamine absortion units (Gary et al. 2007). As the technology matured, more advanced nickel-molybdenum and nickel-tungsten catalysts emerged, capable of addressing more recalcitrant compounds and additional contaminants.

The expansion of hydroprocessing operations created substantial volumes of hydrogen sulfide that required safe and economical disposal, driving development and refinement of the Claus process for sulfur recovery. Modern Claus plants achieve 94–97% sulfur recovery efficiency through two-stage operation: thermal conversion at 1000–1200 °C followed by catalytic conversion at 200–350 °C over alumina-based catalysts. The integration with tail gas treatment technologies enabled refineries to achieve > 99% overall sulfur recovery while meeting stringent SO_2 emission limits, transforming what was initially a waste disposal problem into valuable byproduct recovery that generated marketable sulfur worth $100–300 per ton.

Hydrocracking technology, commercialized in the 1960s, represented a more profound transformation. Unlike hydrotreating, which primarily aimed to remove impurities, hydrocracking enabled the conversion of heavy fractions into lighter, more valuable products. This was achieved through the simultaneous cracking of carbon–carbon bonds and hydrogenation of the resulting fragments, preventing the formation of coke and unsaturated compounds. The development of bifunctional catalysts, combining acidic supports (typically silica-alumina or zeolites) with hydrogenation metals, was crucial to this advancement (Speight 2011).

The evolution of hydroprocessing culminated in the ultra-low sulfur diesel (ULSD) and gasoline technologies developed in the 1990s and 2000s. These processes enabled the production of gasoline and diesel fuels with sulfur content below 10–15 ppm, representing a dramatic improvement over previous capability. This was achieved through enhanced catalysts, optimized reactor designs, and more sophisticated process control strategies (Stanislaus et al. 2010).

The hydroprocessing revolution holds several valuable lessons for net zero refineries. First, it demonstrates the importance of regulatory anticipation. Refiners who invested early in advanced hydroprocessing capacity often gained competitive advantages as environmental regulations tightened. Similarly, early investment in carbon reduction technologies may position refiners advantageously as climate regulations evolve.

Second, the development of hydroprocessing underscores the critical role of hydrogen management in refinery operations. The optimization of hydrogen production, recovery, and utilization became essential skills for refiners, anticipating the challenges and opportunities associated with green and blue hydrogen in net zero refineries (Elsherif, 2015). The hydrogen infrastructure developed for hydroprocessing may serve as a foundation for broader decarbonization efforts.

Finally, the Claus process experience offers valuable lessons for carbon management in net zero refineries, demonstrating how systematic capture and conversion of process emissions can transform environmental liabilities into valuable products. The technical challenges of handling corrosive hydrogen sulfide streams while maintaining high conversion efficiency parallel contemporary challenges in carbon capture and utilization systems. The economic success of sulfur recovery provides a model for developing carbon utilization pathways that create value from captured CO_2 emissions rather than treating them solely as disposal requirements, establishing precedents for byproduct valorization through integrated process design.

3.8 Energy Integration and Process Optimization

Energy integration and process optimization represent perhaps the most significant efficiency improvements in refinery operations over the past half-century. Prior to the 1970s, refineries operated with relatively isolated process units from an energy perspective, with limited heat recovery between processes. Energy consumption was high and thermal efficiency low, with energy viewed as an inevitable operating cost rather than a resource to be optimized (Klemeš et al. 2013).

The oil crises of 1973 and 1979 fundamentally altered this perspective, driving a substantial focus on energy efficiency. The sudden escalation in energy costs heavily damaged the refinery economics, motivating systematic approaches to minimizing consumption. Concurrent advances in applied thermodynamics, particularly the development of pinch analysis methodology by Linnhoff and colleagues, provided analytical tools for comprehensive energy integration (Kemp 2007). Growing computational capabilities enabled the modeling and optimization of increasingly complex systems, while global competition intensified pressure to reduce operating costs.

The implementation of energy integration followed a logical progression. Initial efforts focused on "low-hanging fruit" like obvious waste reduction and basic heat recovery through additional exchangers. This was followed by more systematic heat exchanger network optimization using pinch technology principles. Later developments included advanced cogeneration systems, furnace efficiency improvements, and comprehensive energy management systems incorporating real-time optimization (Kemp 2007).

The quantitative impact of these improvements has been substantial. Studies suggest that the energy intensity of refining operations decreased by approximately 30–40% between 1970 and 2000 in leading facilities, though with significant variation between regions and individual refineries (Worrell and Galitsky 2005).

The energy integration experience provides several critical insights for net zero refinery development. Most fundamentally, it demonstrates the value of systemic approaches to resource optimization. Energy integration succeeded precisely because it transcended unit level improvements to optimize entire facilities, a principle equally applicable to carbon management. Just as heat exchanger networks distribute thermal energy optimally throughout a refinery, carbon networks must efficiently allocate carbon capture, utilization, and storage resources.

The hierarchy of interventions that emerged in energy efficiency first reducing demand, then recovering waste, and finally optimizing generation prefigures an analogous hierarchy for carbon management: avoiding emissions where possible, capturing and utilizing unavoidable emissions, and offsetting remaining emissions. This conceptual framework, shown in Fig. 3.6, provides a structured approach to net zero strategy development.

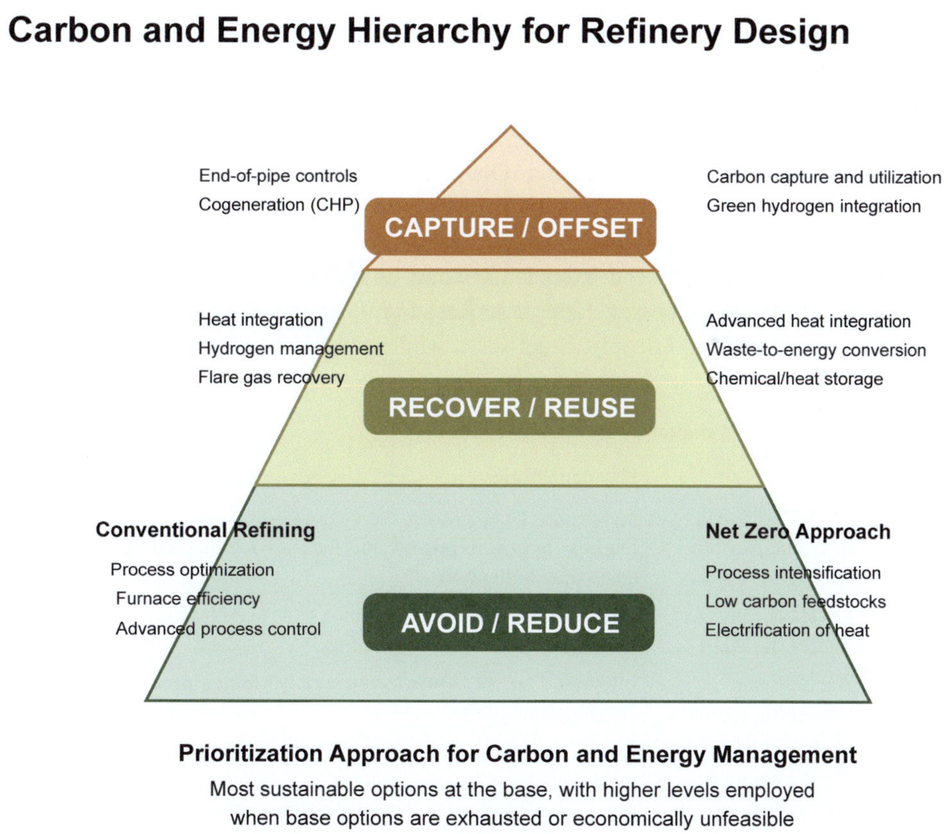

Fig. 3.6 Carbon and energy hierarchy for refinery design

3.9 Digitalization and Automation

The digital transformation of refinery operations represents a profound technological shift in the industry's recent history. Until the 1980s, process control was predominantly analog, with pneumatic systems, basic control valve PID algorithms, and substantial reliance on operators for manual adjustments. Data collection and analysis were limited and primarily oriented toward historical reporting rather than real-time optimization.

This landscape changed due to several facts. The microelectronic revolution made digital processors and systems increasingly affordable and powerful. Competitive economics necessitated margin maximization through operational optimization. Growing process complexity and integration exceeded the capabilities of manual control systems. Advances in algorithms enabled the development of advanced control, optimization, and expert systems (Quelhas et al. 2013).

The progression of refinery digitalization, as shown in Table 3.5, occurred in distinct phases. The initial transition from analog to digital control through Distributed Control Systems (DCS) in the 1980s was followed by implementation of Advanced Process Control (APC) in the 1990s. Subsequent developments included integrated planning and scheduling systems, predictive analytics, and most recently, digital twins enabling comprehensive simulation and optimization (Olaizola, 2022). Process simulators like PROII, HYSYS and ASPEN also provided amazing tools to reproduce real units' performance and dynamics.

The impact of digitalization extended beyond individual unit operations to encompass broader business processes. Enterprise Resource Planning (ERP) systems integrated financial, operational, and supply chain data, while advanced planning tools optimized

Table 3.5 Evolution of digital technologies in refining

Era	Primary technologies	Key capabilities	Limitations
1980s	Distributed control systems	Digital control, improved interfaces, basic data collection	Limited computing power, isolated systems
1990s	Advanced process control, SCADA	Multivariable control, constraint management, remote monitoring	Limited integration, manual optimization
2000s	Planning and scheduling systems, ERP	Integrated business processes, supply chain optimization	Batch processing, limited real-time capability
2010s	Big data analytics, IIoT	Predictive maintenance, pattern recognition, extensive data collection	Data silos, security concerns
2020s	Digital twins, AI/ML, edge computing	Real-time optimization, autonomous operations, advanced simulation	Implementation complexity, skills gap

crude selection, product blending, and distribution decisions. As computational capabilities expanded, refineries developed increasingly sophisticated scheduling systems and supply chain optimization models (Shobrys and White 2002).

Recent advances in artificial intelligence and machine learning have enabled further capabilities. Predictive maintenance systems now anticipate equipment failures before they occur. Virtual sensors estimate difficult to measure parameters based on available data. Computer vision systems monitor for safety issues and process anomalies. These developments represent a progression from algorithmic control to cognitive systems capable of learning and adaptation (Olaizola, 2022).

The digitalization effort offers several important lessons for net zero refinery development. First, the evolution from data scarcity to abundance and its impact on operations, paralleling the role that emissions data must play in carbon management. Comprehensive, real time carbon accounting will be as transformative for emissions reduction as process data was for operational optimization.

Second, the progressive automation of decision making from basic control loops to AI-driven optimization illustrates the pathway for autonomous carbon management systems. The pyramid of automation capabilities that emerged in process control can be adapted to create autonomous carbon optimization systems.

Third, the vertical integration from unit level control to enterprise-wide optimization models the holistic approach required for effective carbon management. Just as modern refineries optimize across the entire value chain rather than individual units, effective decarbonization requires integrated management across process boundaries and potentially across industrial clusters.

3.10 Biorefining and Renewable Process Integration

The integration of renewable feedstocks and processes into conventional refining operations represents one of the most significant recent transformations in the industry. Until the early 2000s, conventional refineries and biofuel plants operated as separate industries. Refineries were designed exclusively for fossil hydrocarbon processing, with no capability to integrate renewable or biogenic feedstocks (Huber et al. 2006).

This separation began to disappear due to a confluence of factors. Renewable fuel policies, including blending mandates and subsidies, created both regulatory pressure and economic incentives for integration. Growing climate awareness increased pressure to reduce the carbon intensity of transportation fuels. Volatile petroleum prices motivated feedstock diversification as a risk management strategy. Advances in biotechnology opened new conversion pathways from biomass to fuel components compatible with existing infrastructure (Serrano-Ruiz et al. 2012).

The evolution of biorefining followed a progression from segregated production toward increasing integration, as shown in Table 3.6. First-generation biofuels like ethanol and fatty acid methyl esters (biodiesel) were produced in dedicated facilities and blended with petroleum products after production. Second-generation approaches began incorporating co-processing of biogenic feeds in conventional refinery units, particularly hydrotreaters. The latest developments include purpose designed processes for converting a wide range of biomass into drop-in hydrocarbons indistinguishable from their fossil counterparts (Karatzos et al. 2014).

Hydroprocessed Esters and Fatty Acids (HEFA) technology, introduced as Hydrotreated Vegetable Oil (HVO) in 2007, represented a significant breakthrough. This process efficiently converted triglycerides into high-quality paraffinic diesel, leveraging modified conventional hydrotreating technology. The resulting renewable diesel was chemically identical to petroleum diesel, eliminating the blending limitations associated with FAME biodiesel (Kubička and Kubičková 2010).

The biorefining experience offers several valuable insights for net zero refinery development. Most notably, it demonstrates the principle of progressive compatibility, the evolution from distinct biofuels requiring separate handling toward drop-in replacements that leverage existing infrastructure. This transition minimized capital requirements and facilitated adoption; a model applicable to carbon management technologies that must similarly integrate with existing assets.

The adaptation of conventional refinery units for biomass processing illustrates the potential for repurposing existing infrastructure during the energy transition. Hydrotreaters

Table 3.6 Biorefining integration approaches in modern refineries

Approach	Feedstocks	Process units	Products	Integration level	Commercial Status
Post-production blending	Corn, sugarcane, vegetable oils	None (separate facilities)	Ethanol, FAME biodiesel	Minimal	Mature
Co-processing	Vegetable oils, used cooking oils	Hydrotreaters, FCC	Renewable diesel, gasoline components	Moderate	Commercial
Dedicated hydroprocessing units	Vegetable oils, animal fats, waste oils	Purpose-built hydrotreating units	Hydrogenated vegetable oil (HVO, SAF)	Moderate-High	Commercial
Biomass gasification	Lignocellulosic biomass, MSW	Gasifiers, syngas processing	Syndiesel, alcohols, hydrogen	High	Demonstration
Catalytic pyrolysis integration	Lignocellulosic biomass	Pyrolysis, hydroprocessing	Bio-oils, drop-in fuels	High	Pilot/ demonstration
Sugar and alcohol conversion	Sugars, alcohols from biomass	Catalytic conversion units	Drop-in aromatics, olefins	High	Emerging

designed for fossil feeds were successfully modified for renewable processing, suggesting that other refinery assets might be repurposed for low-carbon operations.

3.11 Crude-to-Chemicals: Strategic Reorientation of Refining

The most recent transformation in the refining paradigm is the Crude-to-Chemicals (CtC) trend, which represents a fundamental shift in the strategic purpose of petroleum processing facilities. While refineries have historically been configured to maximize transportation fuel production (70–80% of output), CtC technology reorients the objective toward petrochemical feedstock production, significantly increasing their yield from the typical 5–10% to 30–50% or even higher.

This evolution has been driven by several converging factors: forecasts of declining transportation fuel demand due to climate policies and transport electrification, sustained growth in demand for petrochemicals and plastics (particularly in emerging economies), and advances in conversion technologies that enable direct transformation of heavier fractions into chemical feedstocks.

CtC facilities employ highly integrated processing configurations with advanced conversion technologies to maximize the production of light olefins (ethylene, propylene) and aromatics (benzene, toluene, xylenes). This approach represents a fundamental transformation in how a refinery is conceptualized:

1. Extreme Integration: CtC facilities blur the traditional distinction between refinery and petrochemical plant, directly integrating units such as steam cracking, aromatics reforming, and other petrochemical technologies into the main processing chain.
2. Innovative Conversion Technologies: Advanced configurations such as high-severity catalytic cracking processes, selective hydrocracking tailored to maximize olefins, and direct crude-to-olefins (CTO) processes represent specific innovations for this orientation.
3. Atomic Efficiency Maximization: The CtC approach seeks to transfer a greater proportion of carbon atoms from crude oil to value-added chemical products, fundamentally improving resource use efficiency and potentially reducing the carbon footprint per unit of economic value generated.
4. Economic Resilience: By shifting focus toward chemicals markets that are projected to grow even during energy transition, CtC represents a strategic adaptation to maintain refining asset viability in a decarbonizing world.

Several major projects exemplify this trend, including Saudi Aramco's integration with SABIC, Hengli Petrochemical's 20 million tons per annum integrated facility in China, and the Zhejiang Petroleum and Chemical complex, which can convert more than 40% of crude input to chemicals. These facilities represent investments of $20–25 billion each, demonstrating the scale of commitment to this transformation pathway.

The CtC trend offers important insights for net zero transition strategies. It demonstrates how refiners can fundamentally rethink the purpose and configuration of hydrocarbon processing assets in response to changing market and regulatory landscapes. This adaptability, transforming facilities designed for fuel production into chemical manufacturing complexes, suggests similar flexibility might be possible for carbon–neutral conversions. Additionally, the economic logic of shifting to higher-value, longer-lived products rather than combusted fuels parallels aspects of carbon capture and utilization strategies where embedded carbon creates durable value rather than atmospheric emissions.

3.12 Emerging Technologies for Net Zero Refineries

The petroleum refining industry has historically demonstrated remarkable adaptability, evolving from simple distillation operations to increasingly sophisticated conversion processes. The progression from physical separation through thermal and catalytic processing to molecular restructuring follows a logical innovation trajectory.

The transition toward net zero refining requires integration of diverse technologies, shown in Fig. 3.7 and Table 3.7, across preprocessing, conversion, synthesis, and energy systems. Each technology brings specific capabilities while presenting unique challenges for integration.

3.12.1 Advanced Automated Sorting

The development of advanced sorting technologies parallels earlier innovations in petroleum assay and crude characterization, which enabled refiners to process increasingly complex feedstocks. Today's automated sorting systems employ multiple sensor technologies (near-infrared spectroscopy, machine vision, X-ray) with sophisticated algorithms to identify and separate materials from residues used as feedstock, with substantial precision.

These systems have reached commercial maturity (TRL 8–9) with demonstrated capabilities for processing up to 50 tons per hour with > 95% accuracy for most materials (Gundupalli et al. 2017). This represents a significant advancement over previous manual or mechanical sorting methods, enabling the practical utilization of heterogeneous waste streams that were previously uneconomical to process.

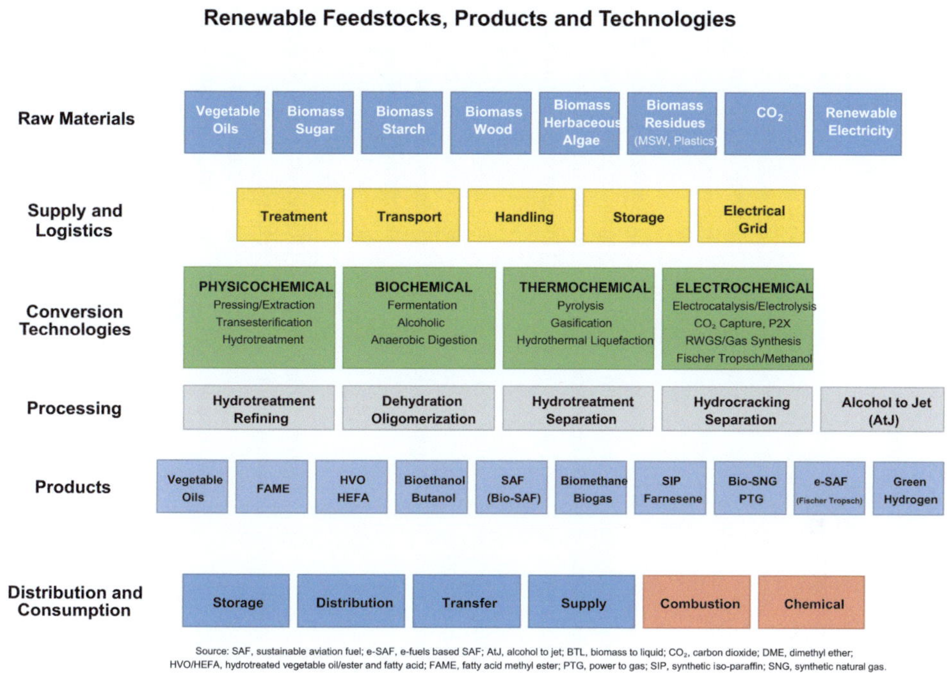

Fig. 3.7 Renewable feedstocks, products and technologies

3.12.2 Catalytic Pyrolysis

Catalytic pyrolysis represents an evolutionary advancement from thermal cracking technologies that changed the early petroleum refineries product slate to gasoline. Just as thermal cracking enabled the conversion of heavy oils into more valuable light products, catalytic pyrolysis facilitates the conversion of waste plastics and biomass into valuable liquid intermediates (Abnisa and Wan Daud, 2014).

Operating at 400–600 °C with transition metal-modified zeolite catalysts, catalytic pyrolysis offers improved product selectivity compared to non-catalytic thermal processes. Current technology status (TRL 6–7) reflects significant progress in pilot facilities processing 50–100 tons/day, but important challenges remain for commercial implementation.

The most critical limitation is catalyst stability and lifetime, which currently ranges from 200 to 1000 h (Iliopoulou et al. 2012), dramatically shorter than the months or years typical for conventional refining catalysts. This creates significant operational and economic challenges that must be addressed through catalyst management strategies or technological advancement.

Table 3.7 Key technologies for net zero refining

Technology	Description	TRL	Primary applications	Key advantages	Limitations	Integration considerations
Advanced automated sorting	Multi-sensor systems with AI algorithms for heterogeneous feedstock separation	08-9	Municipal solid waste, plastic waste separation	> 95% accuracy, up to 50 tons/hour throughput, adaptable to composition changes	Struggles with black materials, mixed-composition items, high humidity conditions	Requires integration with downstream preprocessing, consistent material flow management
Biomass preprocessing	Grinding, drying, torrefaction, and pelletization to standardize feedstock	07-9	Agricultural residues, forestry waste	Improves material consistency, increases energy density, enables reliable feeding	High energy consumption for drying, seasonal variability challenges	Buffer storage needed to manage seasonal availability, heat integration opportunities
Catalytic pyrolysis	Thermal decomposition with catalysts to improve product selectivity	06-7	Waste plastics, biomass	Higher value product slate, lower operating temperatures, improved selectivity	Catalyst deactivation, feedstock sensitivity, complex process control	Requires hydroprocessing of products, catalyst management systems, flexible design for variable feeds
Biomass gasification	Conversion of solid biomass to synthesis gas via partial oxidation	06-8	Agricultural residues, forestry waste, municipal solid waste	Handles diverse feedstocks, produces versatile syngas intermediate	Tar formation, ash management challenges, economy of scale issues	Requires extensive gas cleaning, precise temperature control, material-compatible design

(continued)

Table 3.7 (continued)

Technology	Description	TRL	Primary applications	Key advantages	Limitations	Integration considerations
Fischer–Tropsch synthesis	Catalytic conversion of syngas to liquid hydrocarbons	07-9	Production of paraffinic fuels from syngas	Excellent fuel properties, well-established technology, high selectivity potential	Sensitive to H_2:CO ratio, catalyst poisoning risks, heat management challenges	Heat integration critical, requires high-purity syngas, modular scaling beneficial
CO_2 utilization	Conversion of captured carbon dioxide to valuable products	05-7	Carbon capture integration, negative emissions pathways	Potentially carbon-negative, diverse product pathways	High energy requirements, low conversion rates, limited selectivity	Requires renewable hydrogen sources, catalyst development, process intensification
Advanced hydroprocessing	Hydrogen-based upgrading of intermediate biofuels	06-9	Bio-oil upgrading, renewable diesel and jet production	Removes heteroatoms, improves stability, enhances product properties	High hydrogen consumption, catalyst sensitivity to biocontaminants	Hydrogen management critical, pressure containment design needed, specialized metallurgy
Green hydrogen production	Sustainable hydrogen generation for multiple processes	06-9	Hydroprocessing, CO_2 utilization, fuel cell applications	Zero-carbon hydrogen production, declining cost trajectory	Capital intensive, intermittency challenges with renewables, efficiency limitations	Electricity infrastructure requirements, storage needs, system flexibility for intermittent operation

(continued)

Table 3.7 (continued)

Technology	Description	TRL	Primary applications	Key advantages	Limitations	Integration considerations
Syngas processing	Cleaning and conditioning synthesis gas for downstream use	07-8	Interface between gasification and synthesis	Removes contaminants, adjusts composition, increases system reliability	Energy intensive for compression, complex contaminant management	Critical for system integration, buffer capacity needed, composition monitoring required
Carbon capture	Separation of CO_2 from process and combustion streams	07-8	Emissions reduction, carbon–neutral or negative operation	Multiple technology options, integration with utilization	Energy penalty, solvent/sorbent management, compression needs	Multiple capture points needed, infrastructure for transport, accounting systems
Hydrothermal liquefaction (HTL)	Conversion of wet biomass to biocrude in hot compressed water	05-7	Wet biomass, algae, sewage sludge, food waste	Processes high-moisture (80–90%) feedstocks without drying, lower oxygen content in biocrude	Severe corrosion challenges, high pressure operation (10–25 MPa), complex wastewater treatment	Requires specialized materials, pressure systems, hydroprocessing integration, wastewater management
Solvent liquefaction	Biomass dissolution and conversion in organic solvents	05-6	Lignocellulosic biomass, agricultural residues	Lower operating temperatures (200–350 °C), selective fractionation, reduced char formation	Solvent recovery costs, complex separation systems, limited scale demonstration	Solvent recovery and recycling systems, potential for fractionated product streams

3.12.3 Biomass Gasification

Biomass gasification builds upon decades of coal gasification experience, adapting the technology to handle more reactive but compositionally variable biomass feedstocks. This technology converts solid biomass into synthesis gas through partial oxidation at high temperatures (700–1200 °C).

While gasification has reached commercial maturity (TRL 7–8) for heat and power applications, liquid fuel applications requiring higher gas quality remain at demonstration scale. The primary technical challenge is tar formation, with raw gas containing 5–50 g/Nm^3 of tars (Milne et al. 1998) that can cause equipment fouling and catalyst poisoning in downstream processes.

System availability represents another significant challenge, with current biomass gasification systems achieving 75–85% availability (Kurkela et al. 2016) substantially below the 95+% typical of conventional refining processes. This reliability gap creates significant integration challenges for facilities where downstream processes require continuous, consistent feedstock supply.

3.12.4 Hydrothermal Liquefaction

Hydrothermal liquefaction (HTL) represents an important advancement that builds upon conventional hydroprocessing techniques in petroleum refining. Just as hydroprocessing revolutionized oil refining by incorporating hydrogen to improve product quality, HTL applies similar principles to convert high moisture biomass into biocrude through reactions in hot compressed water.

Operating at 280–380 °C and 10–25 MPa pressure with residence times of 5–30 min, HTL effectively mimics the natural geological processes that formed petroleum but accelerated to minutes rather than millions of years. Current technology status (TRL 5–7) demonstrates promising results in continuous flow demonstration facilities processing 15–50 tons/day, though wider commercial deployment remains limited.

The most significant advantage is HTL's ability to process wet feedstocks (80–90% moisture content) without energy intensive drying, producing biocrude with lower oxygen content (10–20%) than pyrolysis oil (30–40%) and higher energy density (30–36 MJ/kg). However, critical challenges include reactor materials durability under harsh conditions, complex wastewater treatment requirements, and catalyst performance limitations (Peterson et al. 2008). These challenges create substantial engineering and economic hurdles that must be overcome through advanced materials development and process optimization before widespread commercial implementation.

3.12.5 Fischer–Tropsch Synthesis

Fischer–Tropsch (FT) synthesis represents one of the more mature technologies in the net zero refinery portfolio (TRL 7–9), with decades of development for gas-to-liquids (GTL) and coal-to-liquids (CTL) applications. This process converts synthesis gas into long-chain liquid hydrocarbons over metal catalysts, primarily iron or cobalt, at 200–350 °C and 20–40 bar. While the underlying chemistry is well established, adaptation to biomass derived syngas presents significant challenges due to contaminants not typically present in natural gas applications. Catalyst lifetime varies substantially between cobalt (3–5 years) and iron (8–12 months) catalysts under ideal conditions but can be dramatically shortened by biomass derived contaminants (Ail and Dasappa 2016).

From a net zero perspective, the carbon efficiency of 60–75% (Dry 2002) presents additional challenges for maximizing carbon utilization. However, the exothermic nature of FT synthesis (releasing approximately 165 kJ/mol CO) creates valuable opportunities for thermal integration with endothermic processes elsewhere in the facility.

3.12.6 CO_2 Utilization

Carbon dioxide utilization technologies represent the frontier of net zero refining (TRL 5–6), with potential to transform carbon emissions from waste to valuable resource. These technologies parallel the historical development of sulfur recovery systems, which converted problematic hydrogen sulfide emissions into marketable sulfur products.

Multiple conversion pathways exist, including catalytic (e.g., CO_2 to methanol), electrochemical (e.g., CO_2 reduction to syngas), and biological approaches (e.g., algae cultivation). While catalytic conversion to methanol has reached demonstration stage, most pathways remain earlier in development with significant challenges in efficiency and economics.

The fundamental limitation is energy consumption, with current technologies requiring 2–3 times the theoretical minimum energy (Pérez-Fortes et al. 2016). Energy efficiency for CO_2 utilization typically ranges from 35 to 45%, significantly below other conversion processes. Despite these challenges, CO_2 utilization represents a critical technology for achieving truly carbon-negative operation.

3.12.7 Sustainable Hydrogen Production

Hydrogen production represents a critical enabler for multiple net zero refinery processes, from hydroprocessing to CO_2 utilization. While conventional hydrogen production via steam methane reforming dominates current refineries, sustainable hydrogen alternatives are essential for truly carbon neutral operation.

Electrolysis technologies have advanced significantly, with alkaline electrolysis commercially established (TRL 8–9) and PEM electrolysis reaching early commercial stage (TRL 7–8). However, capital costs remain high at approximately \$1000–1800/kW for installed capacity (IRENA 2020), significantly higher than conventional hydrogen production.

The integration of intermittent renewable electricity with continuous hydrogen demand presents additional challenges, requiring substantial buffer storage or hybrid production approaches. This parallels earlier challenges in refinery utility system development, where reliable service had to be maintained despite variable demands and occasional supply disruptions.

Figure 3.8 illustrates the interconnected technologies required for net zero refining, showing material flows (solid lines) and energy inputs (dashed lines) between key process areas. The integration of renewable energy and carbon management is central to achieving carbon neutral or carbon negative operation across the entire refining system.

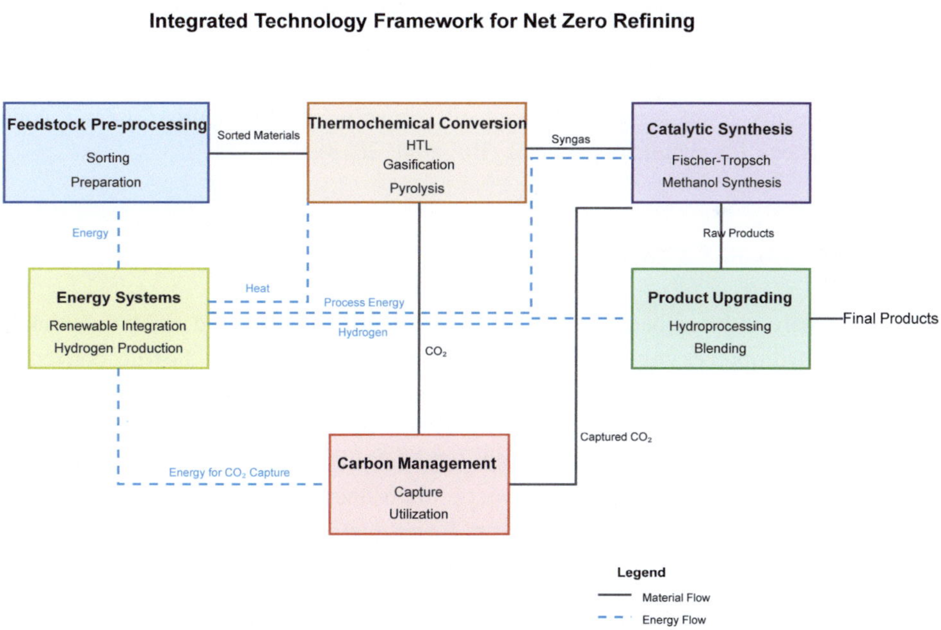

Fig. 3.8 Integrated technology framework for net zero refining

3.13 Innovation Strategies for Net Zero Refineries

The historical analysis of refining technology evolution reveals consistent patterns of innovation which provide insights into potential strategies for developing net zero facilities. Each major advancement in refining technology responded to a combination of external pressures (regulation, market demands, competition) and emerging capabilities (catalysis, digitalization, integration). We have identified the following strategies for net zero refining:

3.13.1 Multidimensional Integration

Multidimensional integration extends beyond traditional energy integration to encompass carbon flows, renewable energy inputs, and cross-sectoral synergies. Just as pinch technology revolutionized energy efficiency, carbon integration methodologies can optimize capture, utilization, and storage resources across industrial clusters.

The utilization principle suggests that all carbon containing streams in a refinery should be viewed as a potential resource rather than waste. This approach extends beyond conventional carbon capture to innovative utilization pathways for different CO_2 concentrations and contexts. Low-concentration streams might feed microalgae cultivation, medium concentration streams could supply enhanced oil recovery, and high-purity streams could serve as feedstock for chemical synthesis (Bui et al. 2018).

3.13.2 Advanced Catalysis

Advanced catalysis has repeatedly revolutionized refining, from fluid catalytic cracking to selective hydroprocessing. Similarly, novel catalysts for carbon dioxide conversion represent one of the most promising paths to economically viable carbon utilization.

Catalytic systems that enable direct conversion of CO_2 to valuable products such as olefins, aromatics, or oxygenates could potentially transform carbon from waste to resource (Artz et al. 2018). Bifunctional or multifunctional catalysts that simultaneously enable multiple reaction pathways can lower energy barriers and improve process economics.

3.13.3 Crisis-Driven Acceleration

Historical evidence indicates that crises accelerate innovation by creating urgency, aligning incentives, and mobilizing resources. The climate crisis can similarly function as an innovation accelerator if properly leveraged. Moonshot programs focused on key enabling technologies such as direct air CO_2 capture (DAC), electrocatalytic CO_2 conversion, or low carbon hydrogen production can compress development timelines through targeted investment and cross-disciplinary collaboration (Mazzucato 2018).

3.13.4 Energy System Integration

The biorefining experience demonstrates the value of progressive compatibility and infrastructure repurposing during transition periods. Rather than wholesale replacement of assets, a phased approach that gradually converts existing units for lower carbon operations can minimize capital requirements while steadily reducing emissions. Strategic investments in flexible capabilities such as units that can process both conventional and bio-feedstocks can facilitate this transition while managing uncertainty.

The combined implementation of these strategies requires a systems approach that avoids optimization silos. Integration across traditional boundaries, between process units, between refineries and adjacent industries, between energy and material flows will be essential. The most successful net zero transformations will likely combine multiple strategies in a coherent framework tailored to specific facility characteristics, regional contexts, and company capabilities.

3.14 Conclusion: The Evolutionary Pathway to Net Zero Refineries

The technological evolution of petroleum refining from simple distillation through increasingly sophisticated conversion and treating processes demonstrates the industry's remarkable capacity for innovation in response to changing market demands, regulatory requirements, and societal expectations. This historical pattern of transformation provides valuable context for approaching the net zero challenge not only as a substantial disruption, but as the next chapter in a continuing story of technological adaptation.

The emerging technologies for net zero refineries from advanced preprocessing and thermochemical conversion to catalytic synthesis and sustainable energy systems represent the logical extension of refining's evolutionary pathway. These technologies build upon established principles while addressing the unique challenges of carbon neutrality (Vogts and Weckhuysen, 2024).

Several key principles emerge from this evolutionary perspective:

1. **Technological innovation in refining has consistently accelerated in response to external pressures**, suggesting that climate imperatives may similarly drive accelerated development of decarbonization technologies.

2. **Major advances have often resulted from technology transfer and adaptation rather than new inventions**, highlighting the importance of cross-sectoral knowledge exchange for net zero transformation.
3. **Integration of energy, materials, and information has consistently delivered performance improvements beyond what individual unit optimizations could achieve**, underscoring the critical importance of system level approaches to decarbonization.
4. **Successful innovations have balanced multiple objectives,** product quality, economic viability, operational reliability, and environmental performance rather than optimizing single dimensions, a principle equally applicable to net zero transformation.

The integration of sustainable technologies with existing infrastructure, the adaptation of historical innovations to new challenges, and the progressive development of increasingly sophisticated carbon management capabilities may represent a viable pathway toward a sustainable refining industry that continues to meet society's needs while addressing climate imperatives.

The technological foundation established in this chapter from traditional refining processes through emerging net zero technologies provides the essential context for the prompt engineering methodology developed in subsequent chapters. This evolutionary perspective on refining technology informs the design approach presented in Chap. 5, where systematic prompt engineering is applied to develop a comprehensive net zero refinery concept that integrates historical experience and knowledge with emerging capabilities.

References

Abnisa F, Wan Daud WMA (2014) A review on co-pyrolysis of biomass: an optional technique to obtain a high-grade pyrolysis oil. Energy Convers Manage 87:71–85

Ail SS, Dasappa S (2016) Biomass to liquid transportation fuel via Fischer Tropsch synthesis—technology review and current scenario. Renew Sustain Energy Rev 58:267–286

Antos GJ, Aitani AM (2004) Catalytic naphtha reforming. Marcel Dekker, New York

Artz J, Müller TE, Thenert K, Kleinekorte J, Meys R, Sternberg A, Bardow A, Leitner W (2018) Sustainable conversion of carbon dioxide: an integrated review of catalysis and life cycle assessment. Chem Rev 118(2):434–504

Bui M, Adjiman CS, Bardow A, Anthony EJ, Boston A, Brown S, Fennell PS, Fuss S, Galindo A, Hackett LA, Hallett JP (2018) Carbon capture and storage (CCS): the way forward. Energy Environ Sci 11(5):1062–1176

Chandler AD (1990) Scale and scope: the dynamics of industrial capitalism. Harvard University Press, Cambridge

Dry ME (2002) The Fischer-Tropsch process: 1950–2000. Catal Today 71(3–4):227–241

Enos JL (1962) Petroleum progress and profits: a history of process innovation. MIT Press, Cambridge

Elsherif M, Mana ZZ, Kamsah MZM (2015) State of the art of hydrogen management in refinery and industrial process plants. J Natural Gas Sci Eng 24:346–356

Forbes RJ (1958) Studies in early petroleum history. E.J. Brill, Leiden
Gary JH, Handwerk GE, Kaiser MJ (2007) Petroleum refining: technology and economics, 5th edn. CRC Press, Boca Raton
Gundupalli SP, Hait S, Thakur A (2017) A review on automated sorting of source-separated municipal solid waste for recycling. Waste Manage 60:56–74
Huber GW, Iborra S, Corma A (2006) Synthesis of transportation fuels from biomass: chemistry, catalysts, and engineering. Chem Rev 106(9):4044–4098
Iliopoulou EF, Stefanidis SD, Kalogiannis KG, Delimitis A, Lappas AA, Triantafyllidis KS (2012) Catalytic upgrading of biomass pyrolysis vapors using transition metal-modified ZSM-5 zeolite. Appl Catal B 127:281–290
IRENA (2020) Green hydrogen cost reduction: scaling up electrolysers to meet the 1.5 °C climate goal. International Renewable Energy Agency, Abu Dhabi
Jones DSJ, Pujadó PR (2006) Handbook of petroleum processing. Springer, Dordrecht
Karatzos S, McMillan JD, Saddler JN (2014) The potential and challenges of drop-in biofuels. IEA Bioenergy Task 39:1–21
Kemp IC (2007) Pinch analysis and process integration: a user guide on process integration for the efficient use of energy, 2nd edn. Butterworth-Heinemann, Oxford
Klemeš JJ, Varbanov PS, Kravanja Z (2013) Recent developments in process integration. Chem Eng Res des 91(10):2037–2053
Kubička D, Kubičková I (2010) Conversion of vegetable oils into transportation fuels. Catal Today 149(1–2):22–31
Kurkela E, Kurkela M, Hiltunen I (2016) Steam–oxygen gasification of forest residues and bark followed by hot gas filtration and catalytic reforming of tars: results of an extended time test. Fuel Process Technol 141:148–158
Mazzucato M (2018) Mission-oriented innovation policies: challenges and opportunities. Ind Corp Chang 27(5):803–815
Meyers RA (2004) Handbook of petroleum refining processes, 3rd edn. McGraw-Hill, New York
Milne TA, Evans RJ, Abatzoglou N (1998) Biomass gasifier "tars": their nature, formation, and conversion. National Renewable Energy Laboratory, Report No. NREL/TP-25357
Moser FX (1996) Catalytic reforming: start-up, operation, regeneration. Hydrocarb Technol Int 4:83–87
Nelson WL (1958) Petroleum refinery engineering, 4th edn. McGraw-Hill, New York
Occelli ML (2010) Fluid catalytic cracking VII: materials, methods and process innovations. Elsevier, Amsterdam
Olaizola IG, Quartulli M, Unzueta M, Goicolea JI, Florez J (2022) Refinery 4.0, a review of the main challenges of the Industry 4.0 Paradigm in Oil&Gas Downstream. Sensors 22(23):9164
Pérez-Fortes M, Schöneberger JC, Boulamanti A, Tzimas E (2016) Methanol synthesis using captured CO_2 as raw material: techno-economic and environmental assessment. Appl Energy 161:718–732
Peterson AA, Vogel F, Lachance RP, Fröling M, Antal MJ Jr, Tester JW (2008) Thermochemical biofuel production in hydrothermal media: a review of sub- and supercritical water technologies. Energy Environ Sci 1(1):32–65
Quelhas AD, Castro JG, Pinto JM (2013) Advanced process control in petroleum refineries: a review. Braz J Chem Eng 30(4):689–702
Robinson PR, Dolbear GE (2006) Hydrotreating and hydrocracking: fundamentals. In: Practical advances in petroleum processing. Springer, New York, pp 177–218
Sadeghbeigi R (2012) Fluid catalytic cracking handbook, 3rd edn. Butterworth-Heinemann, Oxford
Scherzer J, Gruia AJ (1996) Hydrocracking science and technology. Marcel Dekker, New York

Serrano-Ruiz JC, Luque R, Sepúlveda-Escribano A (2012) Transformations of biomass-derived platform molecules: from high added-value chemicals to fuels via aqueous-phase processing. Chem Soc Rev 41(14):5262–5284

Shobrys DE, White DC (2002) Planning, scheduling and control systems: why cannot they work together. Comput Chem Eng 26(2):149–160

Speight JG (2011) The refinery of the future. William Andrew Publishing, Norwich

Stanislaus A, Marafi A, Rana MS (2010) Recent advances in the science and technology of ultra low sulfur diesel (ULSD) production. Catal Today 153(1–2):1–68

Vogt ETC, Weckhuysen BM (2015) Fluid catalytic cracking: recent developments on the grand old lady of zeolite catalysis. Chem Soc Rev 44(20):7342–7370

Vogt ETC, Weckhuysen BM (2024) The refinery of the future. Nature 629(8011):295–306

Worrell E, Galitsky C (2005) Energy efficiency improvement and cost saving opportunities for petroleum refineries. Lawrence Berkeley National Laboratory, LBNL-57260

Yergin D (2011) The quest: energy, security, and the remaking of the modern world. Penguin Press, New York

Principles and Applications of Prompt Engineering for Industrial Systems

4

Abstract

This chapter establishes a comprehensive methodological framework for industrial prompt engineering, examining both the evolutionary development of techniques and their strategic application across the industrial lifecycle. The intersection of Large Language Models (LLMs) with industrial applications creates substantial opportunities for accelerating innovation and problem solving in complex technical domains. However, effectively leveraging LLMs for industrial applications requires systematic approaches to prompt engineering specifically tailored to the unique characteristics of industrial contexts. The chapter presents prompt engineering techniques organized in three evolutionary generations: first-generation foundational approaches including simple prompting and few-shot learning; second-generation reasoning techniques featuring chain of thought and role prompting; and third-generation advanced problem exploration methods including tree of thoughts and self-consistency verification. An integrated strategic framework maps these techniques to industrial project phases from problem definition through implementation planning. The methodology addresses critical challenges including hallucination prevention in technical domains, safety–critical application boundaries, and organizational implementation pathways. Through systematic analysis and practical examples from petroleum refining applications, this chapter demonstrates how structured prompt engineering can transform industrial design processes while maintaining appropriate technical rigor and safety standards.

Rafael Larraz, Prompt Engineering and the Transformation of Petroleum Refining.

R. Larraz, *Prompt Engineering and the Transformation of Petroleum Refining*, Synthesis Lectures on Chemical Engineering and Biochemical Engineering,
https://doi.org/10.1007/978-3-031-99728-0_4

4.1 Introduction: Prompt Engineering for Industrial Applications

The intersection of Large Language Models (LLMs) with the industry creates substantial opportunities for accelerating innovation and problem solving in complex technical domains. This convergence enables more comprehensive exploration of design alternatives, systematic evaluation of options, faster iteration cycles, and enhanced knowledge integration across disciplines.

However, effectively leveraging LLMs for industrial applications requires more than ad hoc interactions, it demands systematic approaches to prompt engineering specifically tailored to the unique characteristics of industrial contexts. This chapter establishes a comprehensive methodological framework for industrial prompt engineering, examining both the evolutionary development of techniques and their strategic application across the industrial lifecycle.

4.1.1 Defining Prompts and Their Role in LLMs

A prompt is a specific input text provided to a Large Language Model to elicit a desired response. Prompts serve as the primary interface between users and language models, functioning as instructions that guide the model's generation process. Unlike traditional programming, which relies on explicit code to define behavior, prompts leverage natural language to direct model outputs through implicit patterns and structures.

The prompt's role in an LLM is multifaceted:

1. **Instruction carrier**: Prompts communicate explicit tasks, specifications, and requirements to the model
2. **Context provider**: They establish relevant background information and knowledge domain for the response
3. **Format definer**: Prompts can specify the desired structure, style, and format of outputs
4. **Reasoning guide**: They can direct the model's cognitive processes, encouraging specific types of analysis or thinking paths
5. **Constraint setter**: Prompts establish boundaries and limitations for appropriate responses.

The effectiveness of a prompt depends on its ability to align the model's statistical prediction capabilities with the user's specific intention. As we will see throughout this chapter, prompt engineering has evolved from simple instruction giving to contextual techniques that strategically leverage the internal mechanisms of language models.

The architectural characteristics of transformer based LLMs, particularly decoder-only models with causal masking as studied in Chap. 2, have several implications for prompt

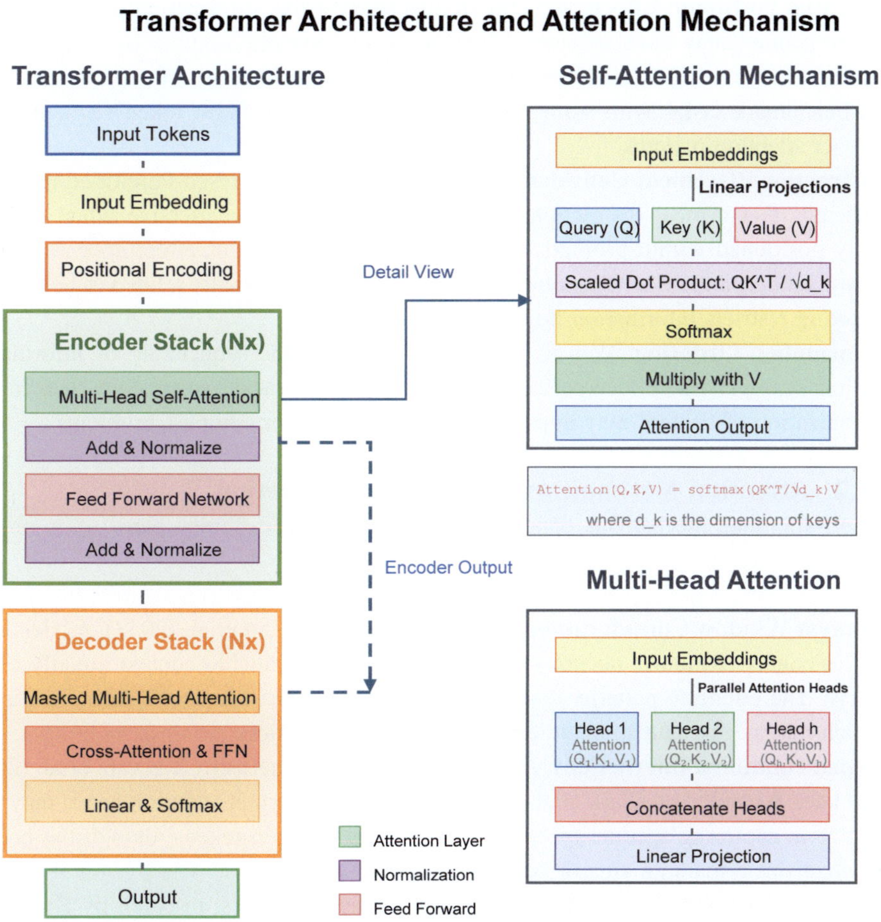

Fig. 4.1 Transformer architecture and key components that influence prompt design

engineering that merit examination. Current research and empirical observations suggest connections between model architecture, as shown in Fig. 4.1, and effective prompting strategies, though the field continues to evolve with ongoing investigation and the results presented rely mainly on empirical analysis.

4.1.1.1 Positional Effects and Information Flow

The causal masking mechanism may create directional effects in how information influences model outputs:

- **Potential Positional Effects**: Several studies suggest that information positioned later in the prompt may have stronger influence on the generated output in many circumstances. Liu et al. (2023) observed that instruction positioning affected performance on benchmark tasks, with some evidence suggesting that latest token sequences can impact attention patterns.
- **Instruction Placement Considerations**: Research by Zamfirescu-Pereira et al. (2023) indicates that critical instructions may sometimes be more effective when strategically positioned within prompts, though the magnitude of this effect appears to vary across models and tasks. This observed behavior may be partially related to how causal masking restricts information flow.
- **Information Ordering**: Wei et al. (2022) noted that the sequencing of information within prompts can affect performance on reasoning tasks, suggesting that careful organization of content may improve model responses for complex problems.

4.1.1.2 Attention Mechanisms and Context Utilization

The self-attention mechanism appears to influence how different parts of a prompt interact:

- **Context Window Considerations**: As demonstrated by Liu et al. (2023), models with larger context windows do not necessarily utilize all available context equally effectively. The attention patterns tend to show varying focus across the token sequence, which may necessitate strategic organization of information.
- **Representational Interactions**: Analysis of attention patterns by Vaswani et al. (2017) and later expanded by Clark et al. (2019) suggests that certain attention heads may specialize in capturing specific types of relationships between tokens, though the precise mechanisms remain an area of active research.
- **Token Economy**: Given the finite nature of context windows, research by Ram et al. (2023) suggests that optimizing token usage based on information importance may improve performance, particularly for complex tasks requiring extensive context.

4.1.1.3 Generation Parameters and Response Characteristics

The autoregressive generation process and its controlling parameters appear to interact with prompt formulation:

- **Temperature-Dependent Prompting**: Empirical observations indicate that different temperature settings may benefit from adjusted prompting strategies. Lower temperatures (0.1–0.5) may work more effectively with precise specifications, while higher temperatures may require more guidance to maintain coherence (Ouyang et al. 2022).
- **Format Specification**: The autoregressive nature of generation suggests that establishing clear response patterns early in the output sequence may influence subsequent token selection. Brown et al. (2020) demonstrated that providing format examples can

effectively guide model outputs, potentially by establishing a structural pattern that subsequent token selection tends to follow.

4.1.2 The Evolutionary Trajectory of Prompt Engineering

Prompt engineering has emerged as a critical discipline for effectively leveraging Large Language Models and its components which are shown in Fig. 4.2. This discipline, as we have seen in the previous section does not rely on a solid theory and has not appeared fully formed, but rather has evolved through distinct generations of techniques, each building upon previous approaches while addressing their limitations.

This chapter examines prompt engineering through an evolutionary lens, tracing its development from simple instructional approaches through increasingly complex methodologies. By understanding this evolution, engineers and technical professionals can select

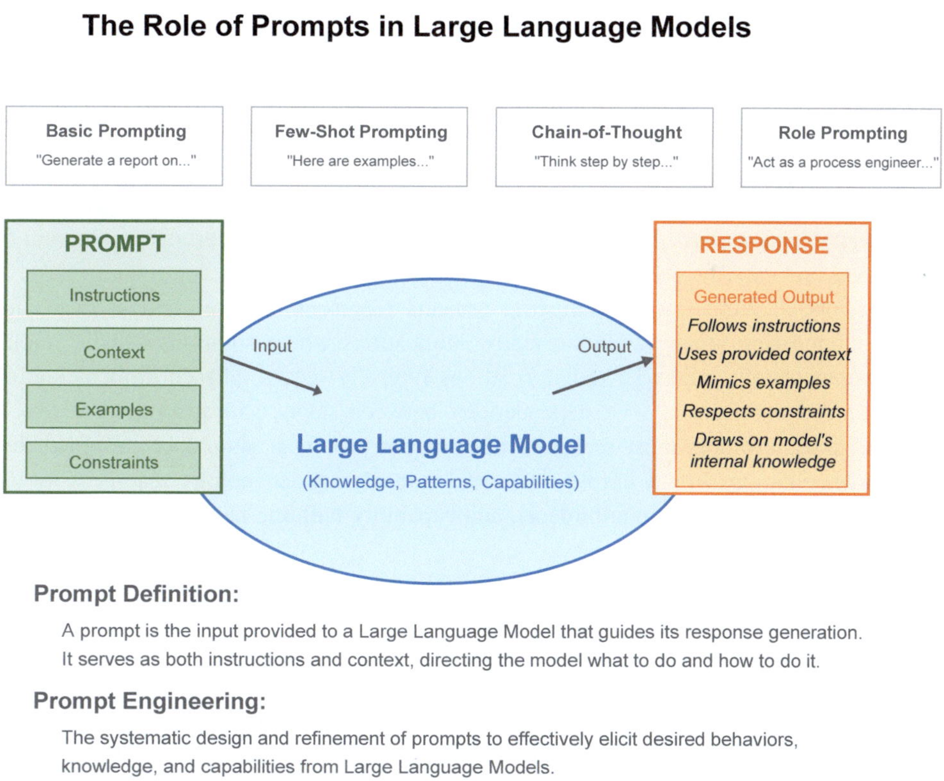

Fig. 4.2 Role of prompt engineering in LLMs

appropriate techniques for specific challenges while appreciating how these methods may continue to develop in response to emerging needs.

4.1.3 Industrial Contexts and Their Unique Requirements

Industrial projects differ fundamentally from many other domains where LLMs are applied, presenting distinct challenges and requirements for effective prompt engineering. These projects typically involve physical systems governed by scientific principles, multidisciplinary integration across technical domains, strict regulatory and safety considerations, and long-term implementation horizons with significant capital investment. This unique context necessitates prompt engineering approaches that differ substantially from those employed in creative writing, general research, or consumer applications.

The technical complexity inherent in industrial projects requires prompts that establish clear system boundaries, define specific physical and operational constraints, and incorporate domain-specific terminology and standards. Industrial systems frequently involve intricate interactions between mechanical, electrical, chemical, and control components, requiring prompts that address both component level details and system level integration challenges.

Industrial projects also operate within strict economic constraints, requiring prompt engineering approaches that balance technical performance with financial viability. The substantial capital investment typically required for industrial implementation means that design decisions have long term consequences, necessitating thorough evaluation of alternatives and careful consideration of implementation pathways.

The temporal dimension of industrial projects represents another critical factor for prompt engineering strategy. Unlike many applications where solutions can be implemented immediately, industrial transformations typically require phased implementation over extended timeframes, with dependencies between stages and adaptation to evolving conditions. Prompts must therefore address not just what should be designed, but how implementation should unfold over time, considering technology readiness levels, infrastructure development requirements, and capability building needs.

4.2 Prompt Engineering Techniques: An Evolutionary Framework

The development of prompt engineering techniques has followed a consistent evolutionary trajectory, with each generation building upon previous approaches while addressing their limitations. This section presents a structured framework that organizes these techniques according to their progression. In Appendix B a complete set of prompt templates and examples is included.

4.2.1 First-Generation Techniques: Fundamental Approaches

The first generation of prompt engineering techniques established fundamental approaches for directing LLM responses. These techniques, while seemingly straightforward, introduced essential principles that still are relevant to more sophisticated methods. First-generation techniques are characterized by their relative simplicity, directness, and focus on basic instruction or example provision.

4.2.1.1 Simple Prompting: Direct Instruction

Simple prompting represents the most basic approach to LLM interaction, involving clear, declarative instructions about the desired response. This technique emerged naturally from early explorations of LLM capabilities and established the foundation for more sophisticated approaches. The effectiveness of simple prompting highlighted the importance of clarity, specificity, and contextual framing principles that remain essential across all prompt engineering techniques.

In industrial contexts, simple prompting proves most valuable for well-defined queries with straightforward answers. For example, requesting technical specifications for common refinery equipment, identifying standard operating parameters, or retrieving established regulatory requirements. The technique's strength lies in its directness and efficiency for questions with clear, factual answers within the model's knowledge domain.

Effective simple prompts for industrial applications typically include:

1. Clear definition of the technical domain
2. Specific question or instruction
3. Desired response format or length
4. Relevant contextual information.

For instance, a simple prompt for hydraulic system troubleshooting might state:

> *As a hydraulic systems specialist, identify the three most common causes of pressure loss in centrifugal pump systems. For each cause, provide the diagnostic indicators and standard remediation approaches. Present your response in a structured format with sections for each cause.*

While effective for straightforward queries, simple prompting shows limitations when addressing complex problems requiring extensive reasoning, complex understanding, or specialized domain knowledge beyond the model's training. These limitations drove the development of more sophisticated techniques that could guide the model's reasoning process more explicitly.

4.2.1.2 One-Shot and Few-Shot Prompting: Learning from Examples

One-shot and few-shot prompting techniques emerged as natural extensions of simple prompting, introducing examples to guide model responses. These approaches leverage the model's pattern recognition capabilities, demonstrating the desired response format and reasoning through representative examples. One-shot prompting provides a single example before requesting a similar response, while few-shot prompting offers multiple examples to establish stronger patterns.

The power of these techniques lies in their ability to establish formats, reasoning patterns, and response structures through demonstration rather than explicit instruction. This proves particularly valuable for industrial applications where specific analysis formats or technical frameworks are expected.

For petroleum refining applications, one-shot and few-shot prompting excel in scenarios requiring standardized analysis formats, such as:

- Equipment failure analysis reports
- Process hazard reviews
- Root cause investigations
- Regulatory compliance assessments
- Management of change evaluations.

A one-shot example for troubleshooting might begin:

> *I need to diagnose operational issues in atmospheric distillation columns in a 100,000 BPD refinery.*

Example:

Input: The distillation column shows an increase in pressure drop from 0.2 psi to 0.5 psi over the last 5 days, while the temperature at tray 15 has increased from 210 to 225 °C. Reflux flow remains constant.
Output:

Diagnosis: The pattern indicates likely flooding in the upper section of the column.

Probable causes:

1. *Excessive column loading (verify feed rate)*
2. *Partial obstruction in the upper trays (possible deposit formation)*
3. *Issue in the condensation system (check condenser performance).*

Recommended actions:—Verify feed rate and reduce if above design capacity—Inspect complete temperature profile to identify exact location of anomaly—Check condensation system and verify subcooling—If persisting, schedule visual inspection at next shutdown.
Now diagnose this issue: *Input: Column overhead temperature has dropped from 145 to 138 °C over 3 days, while bottom temperature increased from 285 to 295 °C. Reboiler steam flow increased 15%.*

The strategic advantage of few-shot prompting increases with the complexity and specificity of the desired response format. By providing multiple examples, the technique establishes both the pattern and its variations, helping the model understand the underlying principles rather than simply mimicking a single instance.

Research has demonstrated that effective examples should:

1. Closely match the complexity of the target problem
2. Represent diverse variations within
3. Represent diverse variations within the pattern
4. Demonstrate the full reasoning process rather than just conclusions
5. Include boundary cases or exceptions where relevant.

Here we present an example of the few-shot technique in an FCC application:

Context: *As a catalysis specialist, I need to recommend adjustments to the catalyst system for an FCC unit processing high metals content feed (nickel: 5 ppm, vanadium: 12 ppm).*

Example 1: *Input: FCC unit with declining conversion (70% to 65%) and increasing dry gas (3% to 4.5%) over the past 4 weeks.*

Output: *Diagnosis: Metal deactivation, primarily nickel catalyzing dehydrogenation reactions. Recommendation: Increase fresh catalyst addition by 20% and implement nickel passivator (antimony) at 3 ppm in the feed. Expected outcome: Conversion stabilization at 68–69% and dry gas reduction to 3.5% within 10–14 days.*

Example 2: *Input: FCC unit with reduced gasoline selectivity (52–48%) and increased coke make (4.5–5.8%) with constant feed.*

Output: *Diagnosis: Deactivation by vanadium contamination affecting zeolite structure. Recommendation: Switch to catalyst with higher rare earth content (from 2 to 3.5%) and add vanadium trap (magnesium oxide) to inventory. Expected outcome: Recovery of selectivity to 50–51% and coke reduction to 5.0–5.2% within 3 weeks.*

Example 3: *Input: FCC unit showing declining gasoline octane (92 to 88 RON) and increasing slurry oil production (8 to 12%) over 6 weeks.*

Output: While powerful for establishing patterns, first generation techniques still faced limitations when addressing problems requiring extensive step by step reasoning or exploration of multiple solution pathways. These challenges led to the development of second-generation techniques focused on explicitly guiding reasoning processes.

4.2.2 Second-Generation Techniques: Reasoning and Specialized Approaches

As applications of LLMs addressed increasingly complex challenges, second generation prompt engineering techniques emerged to guide more challenging and demanding processes. These techniques moved beyond simple instructions or examples to explicitly direct the model's thought process, enabling more reliable handling of multifaceted problems requiring structured analysis or specialized perspectives.

4.2.2.1 Chain of Thought Reasoning: Explicit Analysis Pathways

Chain of Thought (CoT) prompting represents a significant advancement in directing LLM reasoning for complex problems. This technique explicitly instructs the model to break down problems into sequential steps, showing the complete reasoning process rather than simply providing conclusions. By making model performance visible, CoT enables more reliable analysis of multifaceted problems while facilitating verification of the reasoning process.

The strength of CoT prompting lies in its alignment with established problem-solving methodologies in engineering and technical disciplines. Just as engineers decompose complex systems into manageable components, CoT prompting breaks complex reasoning into discrete steps that can be individually assessed and validated.

In petroleum refining applications, CoT prompting proves particularly valuable for:

1. Process troubleshooting sequences
2. Equipment selection decision frameworks
3. Root cause analysis investigations
4. Economic evaluation of project alternatives
5. Safety incident investigations.

A properly structured CoT prompt directs the model to follow a systematic analysis pathway, making each step explicit. For example, a prompt for analyzing catalyst deactivation might request:

> *Analyze potential causes of accelerated catalyst deactivation in a hydrocracker unit. Work through your reasoning step-by-step, considering: 1) Operational parameters that might contribute to deactivation, 2) Feed contaminants that could poison the catalyst, 3) Thermal regime abnormalities, 4) Hydraulic distribution issues, 5) Potential mechanical damage scenarios. For each factor, evaluate the evidence that would support or refute its contribution before moving to the next.*

This structured approach significantly improves reliability for complex analyses by:

- Preventing missing steps in critical reasoning sequences
- Making assumptions explicit and subject to verification
- Allowing detection of logical errors within the reasoning chain
- Creating documented analysis pathways for knowledge transfer.

While CoT substantially improved handling of complex reasoning, it initially required explicit examples demonstrating the desired reasoning approach. This limitation was addressed through the development of Zero-Shot Chain of Thought.

4.2.2.2 Zero-Shot Chain of Thought: Reasoning Without Examples

Zero-Shot Chain of Thought (Zero-Shot CoT) extended the reasoning capabilities of LLMs by enabling structured analysis without requiring specific examples. This technique typically employs simple phrases like "Let's think step by step" or "Let's work through this systematically" to trigger sequential reasoning, leveraging the model's inherent capacity for structured analysis without demonstration.

This advancement proved particularly valuable for industrial applications by:

1. Simplifying prompts while maintaining structured reasoning
2. Enabling consistent analysis approaches across diverse problems
3. Facilitating rapid deployment for novel challenges
4. Maintaining analytical rigor without extensive prompt engineering.

For refining applications, Zero-Shot CoT enables efficient analysis of diverse technical challenges from process optimization to equipment troubleshooting without requiring specialized prompt development for each scenario. This versatility makes it a foundational technique for daily engineering problem solving rather than just specialized applications.

A Zero-Shot CoT prompt for investigating pump vibration might simply state:

A boiler feed water pump is showing increasing vibration levels (5.2 mm/s up from 3.1 mm/s over 14 days) with higher amplitude at 1× running speed. Think step by step to identify the most likely causes, diagnostic approaches to confirm each potential cause, and appropriate remediation strategies.

4.2.2.3 Role Prompting: Specialized Expertise Activation

Role prompting emerged as a powerful technique for accessing specialized knowledge domains within LLMs. This approach instructs the model to assume a specific professional role or expertise when generating responses, effectively narrowing the solution space to a particular domain of knowledge while activating relevant expertise patterns.

The effectiveness of role prompting stems from the model's exposure to domain-specific language, reasoning patterns, and knowledge frameworks during training. By assuming a specific professional role, the model prioritizes relevant knowledge and adopts corresponding analytical approaches, implicitly filtering out irrelevant information.

In petroleum refining applications, role prompting provides access to specialized perspectives including:

1. Process safety specialists for hazard analysis
2. Mechanical engineers for equipment reliability assessment
3. Process engineers for optimization opportunities
4. Instrumentation specialists for control system evaluation
5. Environmental engineers for emissions management.

A well-structured role prompt includes both the professional identity and relevant contextual information about their expertise and approach. For example:

> *As a process optimization specialist with 25 years of experience in refinery operations and particular expertise in heat integration and energy efficiency, review the following heat exchanger network design for a naphtha hydrotreater. Identify potential efficiency improvements in the thermal integration, evaluate the opportunities for heat recovery between process streams, and assess alignment with industry best practices including pinch analysis methodology and minimum approach temperature guidelines.*

The specificity of role definition significantly impacts response quality. More precisely defined roles with clearly articulated expertise areas, methodological approaches, and professional contexts yield more focused and relevant analyses than generic role assignments.

Role prompting can be further enhanced through:

1. Specifying relevant credentials or certifications
2. Indicating years of experience and specialization areas
3. Referencing methodological frameworks used by such professionals
4. Noting industries or organizations where the expertise developed
5. Mentioning regulatory familiarity or compliance knowledge.

This technique complements other approaches and can be effectively combined with CoT reasoning, comparative analysis, and other methodologies for comprehensive problem-solving.

While second-generation techniques significantly advanced LLM reasoning capabilities, they still followed predominantly linear pathways. As industrial challenges, require considering multiple alternatives simultaneously, Third Generation techniques emerged to address these limitations.

4.2.3 Third-Generation Techniques: Advanced Problem Exploration

As industrial applications continued to grow in complexity, third-generation prompt engineering techniques emerged to address multifaceted problems requiring exploration of multiple solution pathways, verification across approaches, and self-improving prompting strategies. These techniques, which represent the current state of the art, enable sophisticated handling of complex design challenges, scenario analysis, and optimization problems.

4.2.3.1 Tree of Thoughts: Branched Analysis for Complex Decisions

Tree of Thoughts (ToT) represents a significant advancement in prompt engineering by enabling exploration of multiple reasoning pathways simultaneously. Unlike linear Chain of Thought approaches, ToT creates a branching structure that can evaluate different possibilities, compare their implications, and select optimal pathways based on assessment criteria.

This approach proves particularly valuable for industrial design challenges that involve:

1. Multiple viable technical approaches
2. Complex trade-offs between competing objectives
3. Uncertain outcomes requiring scenario exploration
4. Interdependent decision points with cascading effects
5. Optimization across multiple parameters simultaneously.

In petroleum refining applications, ToT excels for complex decisions such as:

- Technology selection for major upgrades
- Configuration options for new process units
- Environmental compliance strategy development
- Energy integration opportunity assessment
- Capital project portfolio optimization.

A properly structured ToT prompt explicitly defines both the branches to explore and the evaluation criteria for comparing outcomes. For example, a prompt for evaluating flare gas recovery options might state:

> *Analyze three approaches for flare gas recovery in a 150,000 BPD refinery: 1) Compression to fuel gas system, 2) Liquefaction for LPG recovery, 3) Power generation with waste heat recovery. For each approach, explore: a) Technical feasibility with existing infrastructure, b) Capital and operating cost implications, c) Environmental benefits and regulatory compliance, d) Operational flexibility under varying gas compositions, e) Implementation timeline and complexity. After exploring each branch, compare the approaches based on overall value creation, implementation risk, and alignment with sustainability objectives.*

This branching exploration enables more comprehensive analysis than linear approaches, particularly for decisions with multiple viable alternatives requiring detailed comparison. The explicit mapping of decision branches and evaluation criteria creates transparent documentation of the decision process while ensuring consideration of all relevant factors.

ToT implementation can be enhanced through:

1. Clear definition of each branch to be explored
2. Explicit specification of evaluation criteria
3. Consistent depth of analysis across branches
4. Structured comparison format for alternatives
5. Clear decision recommendation based on assessment.

The primary limitation of ToT approaches involves the complexity of managing multiple branches while maintaining analytical depth. Effective implementation requires careful structuring to ensure each branch receives adequate consideration without overwhelming the analysis.

Figure 4.3 illustrates the concept of a root prompt branching into multiple reasoning paths, each exploring different solution approaches. The tree structure shows how different branches can be evaluated simultaneously, with potential further sub branches for deeper exploration of specific pathways. The visualization demonstrates how ToT enables comprehensive solution space exploration beyond linear reasoning approaches.

4.2.3.2 Self-consistency Techniques: Verification Through Multiple Paths

Self-consistency techniques represent another important third-generation advancement, focusing on verification through multiple independent solution attempts. This approach involves generating several independent solutions to the same problem, then comparing them to identify areas of consensus or disagreement. The principle builds upon established engineering practices where critical calculations are verified through alternative methods or independent checks.

In industrial applications, self-consistency proves particularly valuable for:

1. Safety–critical analyses requiring high confidence
2. Complex calculations with significant impact
3. Regulatory compliance determinations
4. Risk assessment for major investments
5. Technical feasibility evaluations for novel approaches.

The technique can be implemented through explicit prompting for multiple independent analyses. For example:

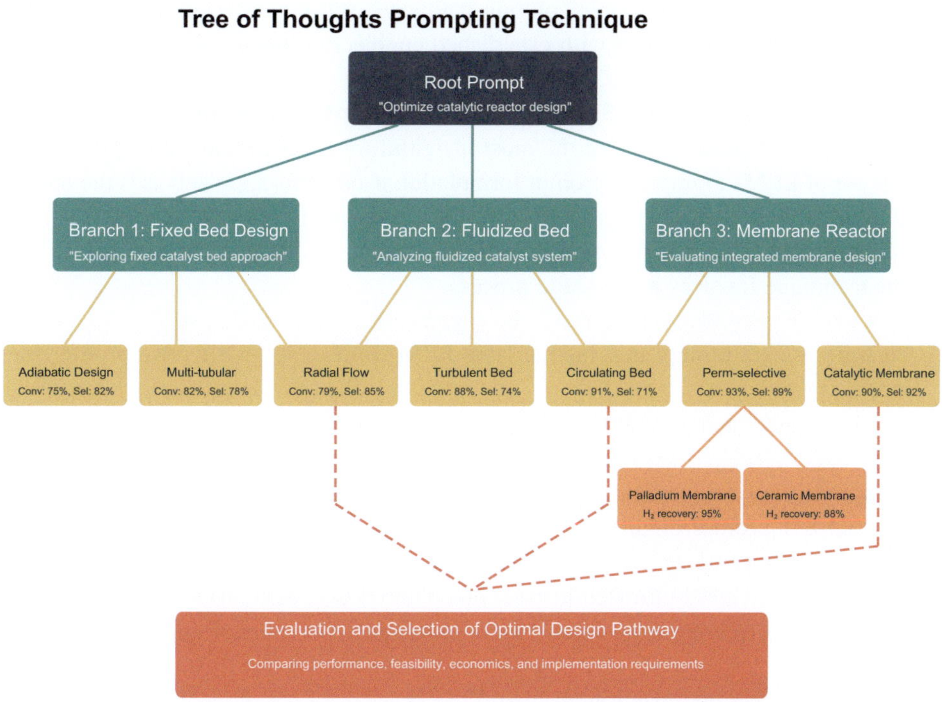

Fig. 4.3 Tree of thoughts approach to solution space exploration

> *Analyze the optimal maintenance interval for a critical heat exchanger in continuous service with fouling tendencies. Generate three independent approaches using: 1) Historical performance degradation analysis, 2) Predictive monitoring based on pressure drop and heat transfer efficiency, and 3) First principles fouling rate modeling. For each approach, show all calculation steps, key assumptions, and economic factors. Then compare the results, identify any significant variances between methods, and recommend a final maintenance schedule with appropriate justification for maximizing run length while minimizing production losses.*

The power of self-consistency lies in its ability to:

- Identify potential errors or oversights in individual approaches
- Reveal differences in assumptions between methods
- Build confidence when multiple methods converge
- Highlight areas requiring further investigation when discrepancies emerge
- Provide documentation of verification for critical decisions.

This approach directly addresses one of the significant challenges in LLM applications for industrial settings: establishing confidence in outputs for consequential decisions. By systematically comparing multiple solution pathways, self-consistency provides a robust verification mechanism aligned with established engineering practices.

4.2.3.3 Meta Prompting: Self-improving Prompt Development

Meta prompting represents one of the most insighful third generation techniques, involving the use of LLMs to improve prompt formulation itself. This approach uses the model's capabilities to refine prompts iteratively, creating increasingly effective instructions based on initial responses or specific requirements.

The technique involves a two-stage process:

1. Creating a meta prompt that instructs the LLM to develop an effective prompt for a specific task
2. Using the generated prompt to address the actual problem.

This approach proves particularly valuable for complex industrial challenges requiring specialized prompt engineering but where expertise in prompt development may be limited. By leveraging the model's understanding of effective prompting patterns, meta-prompting creates optimized instructions tailored to specific technical challenges.

For petroleum refining applications, meta prompting can enhance approaches to:

1. Complex process optimization problems
2. Multidisciplinary design reviews
3. Comprehensive risk assessment frameworks
4. Regulatory compliance analysis
5. Technology evaluation methodologies.

A simple meta-prompting example might begin:

> *Develop a comprehensive prompt that would enable detailed analysis of corrosion mechanisms in crude oil preheat exchangers. The prompt should guide systematic consideration of flow conditions, metallurgy, crude composition factors, temperature profiles, and mitigation strategies. The prompt structure should facilitate thorough analysis while maintaining practical relevance for a process engineer evaluating inspection findings.*

The resulting optimized prompt can then be used to address the actual technical challenge with greater effectiveness than might be achieved through direct prompting without this refinement stage.

Meta-prompting creates several advantages:

- Tailors prompts to specific technical challenges
- Incorporates best practices in prompt engineering automatically
- Reduces dependency on specialized prompt engineering expertise

- Creates documented prompt templates for recurring challenges
- Enables continuous improvement of prompting approaches.

This technique can be particularly valuable in organizations early in their LLM adoption journey, where prompt engineering expertise may be developing but technical domain knowledge is strong. Meta prompting bridges this gap by leveraging the model's understanding of effective prompting patterns.

4.2.3.4 ReAct: Reasoning and Acting with Structured Thinking

ReAct (Reasoning + Acting) represents a sophisticated prompt engineering technique that integrates structured reasoning with planned action steps, enabling LLMs to approach complex problems through an iterative cycle of thought and action. This technique enhances problem solving capabilities for multi-step challenges that require both analytical thinking and concrete actions.

ReAct operates on several fundamental principles:

1. **Interleaved Reasoning and Action**: Alternating between analytical thinking and specific action steps, creating a dynamic problem-solving process
2. **Explicit Thought Verbalization**: Making reasoning processes visible through "Thought" steps that explain the model's analysis before each action
3. **Structured Action Definition**: Clearly defining concrete actions, the model should consider after reasoning through options
4. **Observation Integration**: Incorporating results or feedback from each action into subsequent reasoning cycles

The ReAct technique follows a structured format with distinct components:

1. **Thought**: Explicit reasoning steps that analyze the current situation, evaluate options, or plan next actions
2. **Action**: Specific, concrete steps to be taken based on preceding reasoning
3. **Observation**: Results or information obtained from completed actions
4. **Reflection**: Periodic assessment of progress and strategy adjustment.

A possible template of the ReAct prompt:

> *For each step in solving this problem: 1. Think: Analyze the current state and consider what information you need 2. Act: Determine the most appropriate next action based on your thinking 3. Observe: Note what you learn from the action 4. Reflect: Periodically assess your approach and adjust as needed.*
>
> *Continue this cycle until you reach a complete solution.*

In industrial engineering contexts, ReAct proves particularly valuable for:

- Troubleshooting complex equipment failures
- Developing multi-stage process optimization strategies
- Creating diagnostic decision trees
- Planning sequential maintenance procedures
- Scenario analysis for operational contingencies.

A ReAct prompt example might be:

> *Using the ReAct approach (Reasoning and Acting), develop a systematic process for diagnosing and resolving a refinery pump cavitation issue:*
>
> *Thought: [Analyze what might cause cavitation in refinery pumps] Action: [Determine first diagnostic step to verify cavitation] Observation: [Note what this diagnostic would reveal] Thought: [Based on this observation, reason about likely root causes] Action: [Specify next test or measurement] ...*
>
> *Continue this reasoning and action cycle until you have a complete diagnostic and resolution pathway with decision points for different scenarios.*

This approach enables systematic problem solving that makes each step of the decision process explicit and trackable. It creates a structured trace of both the reasoning and action components, allowing verification of the logical progression and identification of potential gaps or errors.

When applied to industrial challenges, ReAct produces solutions that are not only more systematic but also more implementable, as they explicitly map out the concrete actions required rather than remaining at a conceptual level. The technique proves particularly valuable for knowledge intensive tasks that require both analytical reasoning and practical implementation steps.

4.2.4 Integrated Approaches: Combining Techniques for Maximum Effectiveness

As prompt engineering discipline matures in an organization, the most effective applications in industrial settings increasingly combine multiple techniques into integrated approaches that leverage the strengths of each methodology. These combined approaches create synergistic capabilities exceeding what any single technique could achieve, enabling sophisticated handling of complex industrial challenges.

4.2.4.1 Retrieval Augmented Generation: Integrating Domain-Specific Knowledge

Retrieval Augmented Generation (RAG) represents a powerful integration of LLM capabilities with external knowledge sources. This approach combines the generative capabilities of LLMs with explicit retrieval of relevant information from specialized databases, technical documentation, or organizational knowledge repositories. By providing specific contextual information for each query, RAG addresses one of the fundamental limitations of standalone LLMs: access to current, specialized, or proprietary information beyond their training data. A possible RAG architecture is presented in Fig. 4.4.

In petroleum refining applications, RAG creates particularly valuable capabilities by connecting LLMs to:

1. Equipment specification databases
2. Proprietary process design guidelines
3. Historical incident reports and lessons learned
4. Current regulatory requirements and interpretations
5. Company-specific procedures and standards
6. Operational data and performance trends.

This integration enables LLMs to provide responses grounded in organizational context and current technical information rather than relying solely on general knowledge from their training.

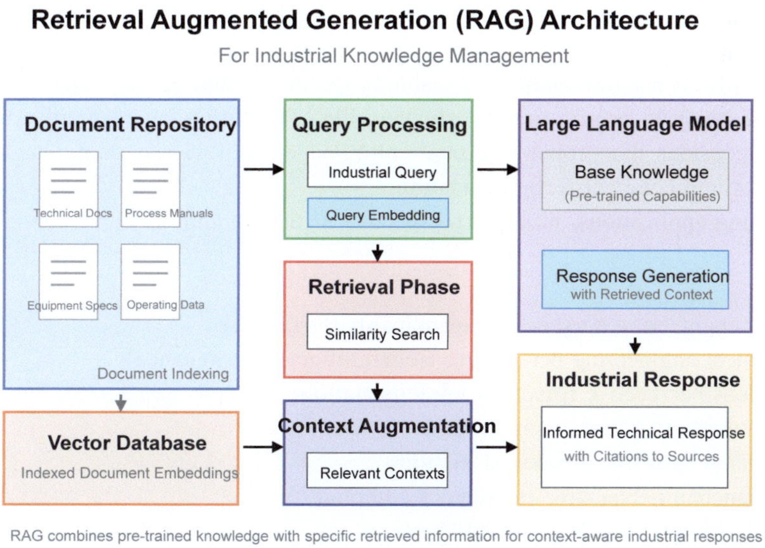

Fig. 4.4 Retrieval augmented generation (RAG) architecture

The implementation of RAG involves:

1. Creating structured knowledge bases of relevant technical information
2. Developing retrieval mechanisms to identify relevant content
3. Designing prompt templates that incorporate retrieved information
4. Establishing verification processes for response accuracy.

A RAG-based approach for equipment troubleshooting might retrieve specific technical manuals, historical maintenance records, and operating procedures relevant to the equipment in question, then incorporate this information into the prompt:

> *Using the attached maintenance history for Pump P-1501A, manufacturer's specification sheet, and recent operating data, analyze the potential causes of the observed bearing temperature increase from 65°C to 78°C over the past week. The pump is a 6-stage vertical turbine pump operating in boiler feed water service as described in the attached documents.*

This integration of specific technical documentation with LLM reasoning capabilities creates responses that combine the generative power of the model with the precision of retrieving exact technical specifications, operating parameters, and historical context.

4.2.4.2 Multiple Perspective Analysis: Integrated Viewpoints

Multiple perspective analysis combines elements of role prompting with structured comparative frameworks to generate integrated analysis incorporating diverse viewpoints. This approach systematically examines challenges from multiple disciplinary or stakeholder perspectives, creating comprehensive understanding beyond what any single viewpoint could provide.

For complex industrial challenges spanning traditional disciplinary boundaries, multi-perspective analysis proves particularly valuable. The technique explicitly directs the model to consider the problem through multiple defined views, then integrate these perspectives into coherent recommendations.

In refining applications, multiple perspective analysis might examine capital projects from:

1. Process engineering perspective (technical performance)
2. Mechanical engineering perspective (reliability and maintenance)
3. Operations perspective (practical implementation)
4. Safety perspective (risk management)
5. Financial perspective (economic viability)
6. Environmental perspective (regulatory compliance)
7. Strategic perspective (long-term alignment).

A comprehensive multiple perspective prompt would direct each viewpoint analysis, specify integration mechanisms, and request balanced recommendations:

> *Analyze the proposed fluidized catalytic cracking unit revamp project from the following perspectives: process engineering, mechanical integrity, operations, safety, financial, environmental, and strategic alignment. For each perspective, identify key considerations, potential concerns, and recommended modifications. Then develop an integrated recommendation that balances these perspectives, identifying critical requirements and potential compromises to optimize overall value.*

This approach creates several advantages for complex industrial problems:

1. Prevents overlooking critical factors outside disciplinary boundaries
2. Identifies potential conflicts between stakeholder priorities
3. Facilitates cross disciplines communication
4. Creates balanced solutions addressing diverse requirements
5. Aligns with natural cross-functional review processes.

The structured integration of multiple perspectives proves particularly valuable for transformation challenges like sustainability initiatives or digital transformation, where success requires alignment across numerous organizational functions and external stakeholders.

4.2.4.3 Context-Rich Framing: Establishing Comprehensive Problem Boundaries

Context rich framing represents a foundational prompt engineering technique that provides detailed background information to enable the LLM to develop responses grounded in specific industrial circumstances rather than generic assumptions. By establishing a comprehensive situational context, this technique ensures that generated content accounts for the full complexity of the industrial environment.

Context-rich framing functions as a prerequisite for effective LLM performance on complex industrial problems by establishing baseline knowledge, environmental conditions, constraints, and objectives. Unlike simpler prompting techniques, context-rich framing provides multidimensional grounding that enables more sophisticated analysis tailored to the specific industrial situation.

Context-rich framing proves particularly valuable in petroleum refining due to the industry's highly contextualized nature, where solutions must account for specific feedstock characteristics, equipment configurations, operating constraints, and business priorities. This technique enables:

1. **Site-Specific Analysis**: Tailoring solutions to the particular configurations, constraints, and challenges of individual refineries rather than generic facilities

2. **Historical Continuity**: Incorporating facility history, previous modifications, and operational experience into the analysis
3. **Integrated System Awareness**: Ensuring recommendations account for integration effects across interconnected refinery units
4. **Constraint-Aware Solutions**: Developing practicable solutions that respect physical, regulatory, and economic limitations

The effectiveness of context rich framing increases with the complexity of the industrial problem and the specificity of the contextual information provided. For complex refinery optimization challenges, establishing comprehensive context is not merely helpful but essential for generating implementable recommendations.

A comprehensive context-rich framing example:

> I *need to analyze feasible energy efficiency improvements for a 120,000 BPD crude distillation unit (CDU) at a Gulf Coast refinery with the following characteristics:*
>
> *Operational context: - Current energy intensity: 1.2 MBtu/bbl (vs. industry benchmark of 0.85-1.0 MBtu/bbl) - Original design: 1985, last major revamp: 2008 (preheat train expansion) - Currently processing opportunity crudes with API gravity range 18-24° - Located in a non-attainment zone for NOx emissions - Minimum 96% annual on-stream factor requirement*
>
> *System configuration: - 6-panel preheat train (12 shell-and-tube exchangers, mixed counter/cross flow) - Single fired heater with 4 passes, conventional burners (1985 design) - Atmospheric column with 4 pumparounds and 5 product draws - Vacuum unit with 3 product streams - Existing air preheater for combustion air (70% efficiency) - Limited plot space on north and east sides of unit*
>
> *Business environment: - Capital budget constraint: $15 million maximum - Required ROI: minimum 25% IRR - Carbon pricing anticipated within 3 years ($45/ton initial price) - Corporate mandate to reduce site emissions by 30% within 5 years - Upcoming turnaround scheduled in 14 months with 30-day duration*
>
> *Given this context, identify and prioritize the five most promising energy efficiency improvements for this CDU, including technical approaches, estimated energy savings, implementation requirements, capital costs, and payback periods. Focus particularly on options that can be implemented during the upcoming turnaround and provide significant progress toward the corporate emissions reduction goal.*

The detailed analysis would identify opportunities such as preheat train reconfiguration, advanced combustion controls, furnace air preheater enhancement, pumparound duty optimization, and vacuum system upgrades. Each recommendation would be specifically tailored to the described context accounting for the equipment, spatial constraints, processing requirements, economic parameters, and implementation timeline resulting in actionable, context appropriate solutions rather than generic efficiency measures.

This example demonstrates how context rich framing establishes comprehensive problem boundaries that enable the LLM to generate highly contextualized, practical recommendations that address the specific challenges and constraints of the described facility, while remaining aligned with business objectives and implementation realities.

4.2.4.4 Visualization Prompting: Enhancing Understanding Through Visual Representation

Visualization prompting directs LLMs to conceptualize and describe visual representations of complex systems or processes, bridging textual reasoning with visual thinking. This technique proves valuable for communicating multidimensional relationships that are challenging to comprehend through text alone.

Visualization prompting operates on several key principles:

1. **Spatial Reasoning**: Representing relationships between elements in spatial terms to clarify interactions
2. **Information Density Optimization**: Compressing complex information into comprehensible patterns
3. **Process Sequencing**: Using visual flow representations to clarify sequential operations
4. **Hierarchical Organization**: Structuring information visually to reveal dependencies and organizational principles

The technique can be implemented through several complementary methods:

1. **Descriptive Visualization**: Directing the LLM to describe visual elements in detail, including spatial arrangements and connections
2. **Structural Framework Specification**: Guiding the LLM to articulate specific visual frameworks like process flow schemes
3. **Visual Development Guidance**: Providing specifications for human designers to develop effective visualizations

In industrial engineering contexts, visualization prompting is particularly valuable for:

- Process integration analysis
- System architecture development
- Risk factor mapping
- Performance dashboard design.

A possible implementation of visual prompting is the following example:

Design a visualization of heat integration opportunities between the following refinery process units:

Based on the identified stream data with associated duties and temperature ranges: - Hot Stream 1: Atmospheric column overhead (130°C → 45°C, 4.5 Mcal/h) - Hot Stream 2: FCC main fractionator overhead (145°C → 40°C, 6.0 Mcal/h) - Hot Stream 3: Hydrocracker reactor effluent (380°C → 170°C, 10.1 Mcal/h) - Cold Stream 1: Crude preheat train (30°C → 260°C, 15.5 Mcal/h) - Cold Stream 2: FCC feed preheat (210°C → 340°C, 6.7 Mcal/h) - Cold Stream 3: Boiler feedwater (95°C → 210°C, 3.6 Mcal/h)

CC*reate: 1. A temperature-enthalpy composite curve diagram with clear axis labels and temperature ranges 2. Color coding to distinguish heat availability in different temperature regions (high: >300°C, medium: 150-300°C, low: <150°C) 3. Clear identification of the pinch point with annotations explaining its significance for design 4. Proposed heat exchanger network connections overlaid on a simplified process flow diagram 5. A grand composite curve showing utility requirements*

For each visual element, explain: - The key insights it reveals about energy recovery potential - How it guides practical engineering decisions - Quantified benefits (estimated energy savings, reduced CO_2 emissions, etc.) - Implementation considerations including physical distance between units

Use professional engineering visualization standards and ensure the graphics would be clear to both technical and non-technical personnel.

This approach enables engineers to conceptualize complex relationships more effectively than through narrative text alone, particularly for systems with multiple interacting components or processes with parallel operations.

When combined with other techniques like Chain of Thought, visualization prompting enhances communication of complex concepts and facilitates shared understanding among diverse stakeholders.

4.2.4.5 Domain Integrated Frameworks: Industry-Specific Approaches

The most sophisticated integrated approaches combine multiple prompt engineering techniques within domain-specific frameworks tailored to particular industrial contexts. These approaches embed industry knowledge, standard methodologies, and specialized terminology directly into the prompting structure, creating powerful capabilities for specific application areas.

For petroleum refining applications, domain integrated frameworks might include:

1. Process safety review frameworks combining hazard identification techniques with consequence analysis
2. Technology selection methodologies integrating techno-economic assessment with implementation planning
3. Risk-based inspection approaches combining mechanical integrity analysis with operating context
4. Energy optimization frameworks integrating process integration techniques with economic assessment

5. Environmental compliance strategies combining regulatory analysis with technological solutions.

These frameworks integrate appropriate prompt engineering techniques for each component of the analysis within a coherent overall structure aligned with industry methodologies. For example, a comprehensive process optimization framework might combine:

- Role prompting for process engineering specialist perspective
- Tree of thoughts for efficiency improvement scenario exploration
- Chain of thought for economic impact analysis
- Self-consistency for performance projection verification
- Multi-perspective analysis for operational feasibility evaluation.

The resulting integrated framework creates sophisticated capabilities tailored to specific industrial applications while maintaining alignment with established industry methodologies.

In Fig. 4.5, a matrix visualization displays the effectiveness of different prompt engineering techniques (Simple Prompting, Few-shot Learning, Chain of Thought, Role Prompting, Tree of Thoughts) across various dimensions, it highlights the relative strengths of each technique for different applications, providing a quick reference for technique selection based on problem characteristics.

4.2.5 Safety-Critical Applications: Boundaries and Safeguards

While prompt engineering offers significant potential for accelerating industrial innovation, its application in safety critical contexts requires special consideration and structured safeguards. This section establishes clear boundaries for the application of LLM assisted design in safety domains and provides a framework for responsible implementation.

4.2.5.1 Safety Criticality Classification

The application of prompt engineering techniques in industrial settings should be guided by a systematic classification of safety criticality. The following three-tier framework defines appropriate boundaries:

Tier 1: Ideation and Knowledge Synthesis (Permitted)

- Problem identification and hazard brainstorming
- Compilation and organization of safety regulations and standards
- Synthesis of incident investigation methodologies
- Generation of inspection checklists and protocols

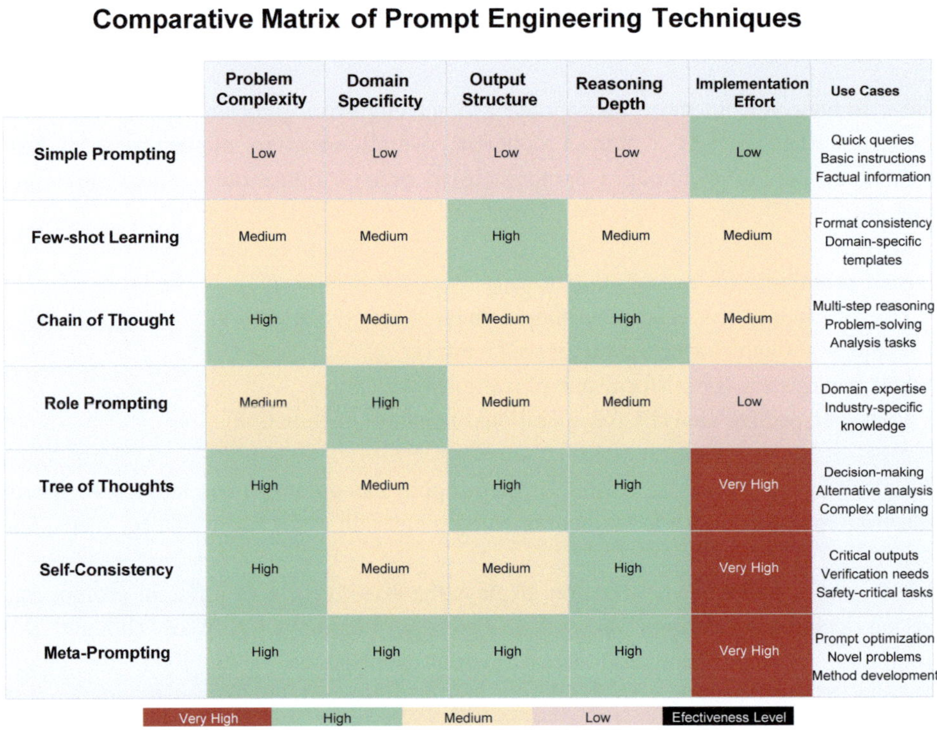

Comparative Matrix of Prompt Engineering Techniques

	Problem Complexity	Domain Specificity	Output Structure	Reasoning Depth	Implementation Effort	Use Cases
Simple Prompting	Low	Low	Low	Low	Low	Quick queries Basic instructions Factual information
Few-shot Learning	Medium	Medium	High	Medium	Medium	Format consistency Domain-specific templates
Chain of Thought	High	Medium	Medium	High	Medium	Multi-step reasoning Problem-solving Analysis tasks
Role Prompting	Medium	High	Medium	Medium	Low	Domain expertise Industry-specific knowledge
Tree of Thoughts	High	Medium	High	High	Very High	Decision-making Alternative analysis Complex planning
Self-Consistency	High	Medium	Medium	High	Very High	Critical outputs Verification needs Safety-critical tasks
Meta-Prompting	High	High	High	High	Very High	Prompt optimization Novel problems Method development

Very High High Medium Low Efectiveness Level

Fig. 4.5 Comparative effectiveness of prompt engineering techniques

- Documentation of safety learnings and best practices.

Tier 2: Analysis with Verification (Restricted)

- Hazard and operability (HAZOP) studies with explicit verification
- Safety procedure development with expert review
- Risk assessment with independent validation
- Investigation of safety incidents with supervised analysis
- Failure mode identification with cross-discipline verification.

Tier 3: Critical Safety Decisions (Prohibited)

- Safety instrumented system design and specification
- Pressure vessel and relief system design parameters
- Critical operating limits and safeguards
- Hazardous area classification
- Emergency response system design.

For Tier 1 applications, standard prompt engineering techniques can be applied with normal quality checks. Tier 2 applications require enhanced verification protocols and should never proceed without independent expert validation. Tier 3 applications should not rely on LLM outputs for decision making; while LLMs may provide supporting information, final determinations must follow traditional engineering methods with appropriate professional oversight.

4.2.5.2 Safety-Enhanced Prompt Engineering Techniques

When applying prompt engineering in safety-adjacent contexts, these enhanced approaches should be considered:

Explicit Limitation Prompting

This technique explicitly directs the LLM to identify knowledge limitations and uncertainties:

As a process safety specialist analyzing [specific system], provide your assessment of [safety concern]. For each aspect of your analysis, explicitly classify your confidence level as: 1. High confidence: Based on well-established principles with minimal uncertainty, 2. Moderate confidence: Based on general principles but with contextual variables, 3. Low confidence: Areas requiring additional data or specialist input. For any analysis where confidence is not high, specify what additional information or expertise would be required for a definitive assessment.

Multi-model Verification

This approach enhance safety assessments using multiple prompt techniques:

Analyze the following safety concern from three different perspectives: 1. First, use a systematic Chain of Thought approach to evaluate [safety concern] step-by-step based on first principles. 2. Next, apply an independent analysis using the Tree of Thoughts method to explore different safety scenarios and their implications. 3. Finally, use a role-based approach as an experienced safety auditor to critically evaluate the previous analyses. Explicitly identify any contradictions or inconsistencies between these approaches and recommend the most conservative pathway forward.

Negative Consequence Exploration

This technique deliberately explores failure modes and potential negative outcomes:

For the proposed [system/approach], conduct a thorough analysis of potential failure modes: 1. Identify at least [N] distinct ways this system could fail, including non-obvious failure modes, 2. For each failure mode, analyze: - Initiating events and contributing factors, - Warning signs and detection methods, - Potential consequences and their severity, - Existing safeguards and their limitations, - Required additional protective measures. Prioritize comprehensiveness over optimism in your analysis, and highlight any scenarios requiring specialized risk assessment techniques.

4.2.5.3 Integration with Process Safety Management Systems

Prompt engineering applications must integrate with established Process Safety Management (PSM) systems rather than bypass them. Key integration points include:

1. **Management of Change (MOC)**: All LLM generated content used in safety contexts must be processed through formal MOC procedures with clear documentation of:
   - The specific prompts used to generate the content
   - Verification steps performed
   - Expert reviews conducted
   - Modifications made to LLM outputs.
2. **Process Hazard Analysis (PHA)**: LLMs may support PHA processes for idea generation, but should not replace structured methodologies such as HAZOP or LOPA. LLM contributions should be clearly identified as inputs requiring validation rather than conclusions.
3. **Documentation and Traceability**: Organizations must maintain complete records of:
   - Original prompts used for safety applications
   - Unmodified LLM responses
   - Verification steps performed
   - Names and qualifications of human reviewers
   - Final approved content with justifications for any modifications.

4.2.5.4 Verification and Validation Requirements

The following verification framework should be applied to any LLM outputs related to safety:

Technical Accuracy

- Verification against established industry codes and standards
- Confirmation of calculations through independent means
- Cross-reference with manufacturer specifications and data sheets
- Compliance with regulatory requirements.

Comprehensiveness

- Verification that all relevant hazard scenarios are considered
- Confirmation that analysis boundaries are appropriately defined
- Assessment of whether risk factors are adequately addressed
- Evaluation of whether edge cases and abnormal conditions are included.

Consistency

- Internal consistency of the analysis and recommendations
- Consistency with established company standards and procedures
- Alignment with industry best practices
- Temporal consistency across operating conditions and lifecycle phases.

Prudence

- Verification that uncertainty is handled conservatively
- Confirmation that protective measures exceed minimum requirements
- Assessment of whether sufficient safety margins are incorporated
- Evaluation of the analysis against worst-case scenarios.

4.2.5.5 Human Expertise Requirements

Effective integration of prompt engineering in safety contexts requires specific human expertise. Organizations should ensure:

1. **Safety Expertise**: Involvement of certified safety professionals with domain-specific experience
2. **Prompt Engineering Specialists**: Team members trained in both prompt engineering and safety disciplines
3. **Independent Verification**: Separation between prompt developers and safety verifiers to maintain independence
4. **Executive Accountability**: Clear responsibility at leadership level for safety governance of AI applications.

4.2.5.6 Practical Implementation

Organizations implementing prompt engineering in safety-adjacent areas should:

1. Start with clearly non critical applications to develop experience and verification protocols
2. Establish formal governance structures before expanding to more sensitive applications
3. Develop competency models specific to AI-assisted safety analysis
4. Create audit mechanisms to ensure compliance with established boundaries
5. Implement anonymous reporting systems for concerns about inappropriate AI application in safety contexts
6. Regularly review and update boundaries as technology and organizational capability evolve.

By maintaining these boundaries and implementing structured safeguards, organizations can responsibly leverage the benefits of prompt engineering while preserving the integrity of safety–critical systems and decisions. This balanced approach enables acceleration of innovation in appropriate domains while ensuring that safety remains paramount in industrial operations.

4.3 Strategic Framework for Industrial Applications

Developing an effective prompt engineering strategy for industrial applications requires a structured approach that aligns with the phases of industrial projects while leveraging the techniques discussed in Sect. 4.2. This strategic framework organizes prompt engineering into a coherent progression that moves from problem definition and conceptual design through solution development to implementation planning, with each phase requiring distinct prompt engineering approaches tailored to its specific cognitive demands.

As shown in Fig. 4.6, each design phase employs specific prompt engineering techniques most suited to its cognitive demands. This systematic application ensures that each phase of the design process benefits from prompt engineering techniques appropriate to its specific challenges.

The effectiveness of this strategic framework relies not just on selecting appropriate techniques for each phase, but on developing a coherent progression where insights from each stage inform subsequent prompt development. This creates a cumulative knowledge-building process where LLM interactions build upon one another, progressively refining the industrial solution.

4.3.1 Problem Definition Phase: Establishing Foundations

The problem definition phase represents the basis upon which subsequent design and implementation activities will be built. In industrial contexts, this phase involves establishing clear system boundaries, mapping technical and operational constraints, identifying stakeholder requirements, and articulating performance objectives. Effective prompt engineering during this phase creates the conditions for successful subsequent development by ensuring comprehensive problem understanding before solution generation begins.

Context-Rich Framing

As described in Sect. 4.2.4.3, context-rich framing provides detailed background information about the industrial context. This includes existing systems, operational environment, regulatory requirements, and historical approaches to similar challenges. By establishing this rich contextual foundation, LLMs can generate responses that account for the specific industrial setting rather than offering generic solutions.

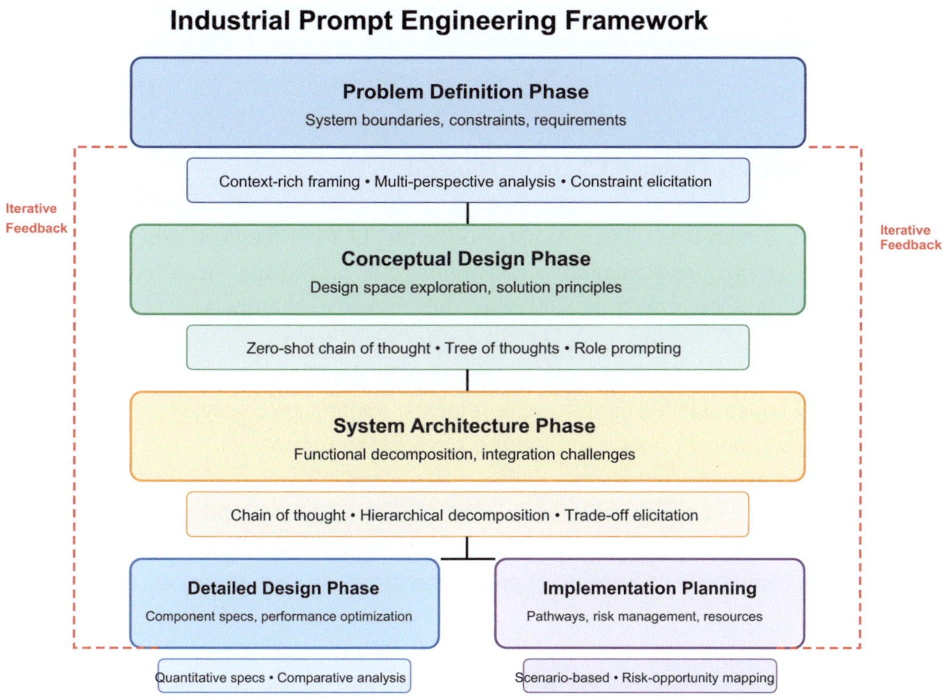

Fig. 4.6 Strategic framework for industrial prompt engineering

For example, a context-rich frame for a refinery project might include:

- Current production capacity and configuration
- Specific equipment types and conditions
- Regulatory environment and compliance history
- Previous similar projects and their outcomes
- Market conditions and competitive pressures.

Multi-perspective Analysis

This application of the Multiple Perspective Analysis technique (Sect. 4.2.4.2) expands on role prompting by examining challenges from multiple viewpoints simultaneously. By considering engineering, operations, maintenance, regulatory, financial, and environmental perspectives, this technique ensures comprehensive problem understanding, particularly for identifying non-obvious constraints and requirements.

A multi-perspective prompt might begin:

Analyze this refinery optimization challenge from the perspectives of: process engineering, operations, maintenance, environmental compliance, business strategy, and community relations...

Constraint Mapping Through Chain of Thought

This application of Chain of Thought (Sect. 4.2.2.1) systematically identifies constraints across multiple dimensions. The prompt guides the LLM through a structured reasoning process to identify and categorize constraints that will shape the solution space. In industrial contexts, these constraints typically fall into several categories:

- Physical constraints (materials, space, temperature limits)
- Operational constraints (reliability requirements, maintenance access)
- Economic constraints (capital limitations, payback requirements)
- Regulatory constraints (emissions limits, safety standards)
- Implementation constraints (shutdown windows, resource availability).

A systematic prompt using Chain of Thought to map constraints might ask the LLM to:

Think step-by-step to identify all constraints that would affect this refinery upgrade project, considering physical limitations, operational requirements, economic boundaries, regulatory compliance, and implementation factors.

The systematic application of these problem definition techniques creates a robust foundation that incorporates diverse perspectives, maps constraints across multiple dimensions, and establishes clear evaluation criteria for subsequent design activities.

4.3.2 Conceptual Design Phase: Exploring the Solution Space

The conceptual design phase in industrial projects focuses on exploring the solution space before detailed development begins, identifying promising approaches that warrant further refinement. This phase requires broad exploration across potential solution principles, technologies, and architectural approaches without premature commitment to specific implementations. Effective prompt engineering during this phase facilitates creative ideation while maintaining grounding in technical feasibility and alignment with defined constraints.

Zero-Shot Chain of Thought

As described in Sect. 4.2.2.2, zero-shot chain of thought enables structured reasoning without requiring specific examples. This technique encourages LLMs to explore solution spaces systematically, considering multiple alternatives before reaching conclusions. For industrial applications, this approach might involve requesting step-by-step consideration of different technology classes, working principles, or architectural paradigms that could address the defined problem.

Few-Shot Prompting

Building on the technique described in Sect. 4.2.1.2, few-shot prompting provides examples that establish patterns for the LLM to follow. In conceptual design, these examples might demonstrate format, depth of analysis, or reasoning patterns that should be applied to the new design challenge. For industrial applications, providing examples of successful concept development creates a template for generating new concepts with similar analytical rigor.

Tree of Thoughts

Building on the technique discussed in Sect. 4.2.3.1, the tree of thoughts approach enables exploration of multiple conceptual branches simultaneously. This approach proves especially effective for industrial challenges where multiple viable paths exist, each with different implications for performance, cost, and implementation. By explicitly mapping these alternative pathways and their potential outcomes, this technique provides a comprehensive view of the solution landscape.

Role Prompting

As detailed in Sect. 4.2.2.3, role prompting directs the LLM to adopt the perspective of specific technical specialists. By instructing the model to respond as process engineers, mechanical designers, control system specialists, or environmental experts, this technique generates diverse solution concepts that leverage domain-specific knowledge and approaches. The integration of these perspectives often yields insights that would not emerge from a single viewpoint.

The combination of these conceptual design techniques creates a comprehensive exploration of the solution space, identifying promising directions for further development while maintaining alignment with project constraints and objectives.

4.3.3 System Architecture Phase: Establishing Structural Foundations

The system architecture phase in industrial projects involves establishing the fundamental structure of the solution, including major subsystems, functional allocations, process flows, and integration approaches. This phase creates the structural framework upon which detailed design will subsequently build, making architectural decisions that will significantly influence system performance, implementation feasibility, and future adaptability.

Chain of Thought Reasoning

Building on the technique described in Sect. 4.2.2.1, chain of thought prompting enables systematic reasoning about system structure, functional allocation, and integration challenges. This technique guides LLMs through the logical development of architectural concepts, considering dependencies, interfaces, and emergent properties that arise from component interactions. For industrial applications, this often involves mapping material flows, energy transfers, information exchanges, and control relationships that define the system architecture.

A Chain of Thought prompt for system architecture might begin:

Let's develop the architecture for this refinery unit step-by-step, first identifying major subsystems, then defining interfaces between these subsystems, next establishing control relationships, and finally evaluating the integration of the complete system.

Comparative Analysis

This application of techniques from Sect. 4.2.1 (enhanced with structured criteria) systematically compares potential architectural approaches. By evaluating alternatives against consistent criteria, this approach identifies relative strengths and weaknesses of different system structures. For industrial applications, architectural comparisons typically address technical performance, reliability, maintainability, scalability, and implementation complexity.

Self-consistency

As described in Sect. 4.2.3.2, self-consistency involves generating multiple independent analyses of the same problem, then comparing them for common elements. For architectural development, this technique can verify the robustness of proposed structures by assessing whether different analytical approaches converge on similar architectural principles. This verification is particularly valuable for complex industrial systems where single-path analysis might miss critical considerations.

Tree of Thoughts for Architecture Exploration

Applying the Tree of Thoughts technique (Sect. 4.2.3.1) to architectural exploration enables systematic consideration of different structural approaches and their implications. By branching the analysis into multiple architectural paradigms and evaluating each branch's strengths and limitations, this technique creates a comprehensive exploration of potential system structures before commitment to a specific approach.

4.3.4 Detailed Design Phase: Refining and Optimizing

The detailed design phase in industrial projects involves transforming architectural concepts into specific technical specifications, including component selection, parameter optimization, material choices, and detailed performance analysis. This phase requires narrower but deeper technical focus than earlier phases, diving into specific engineering domains with greater precision.

Comparative Analysis with Detailed Criteria

Extending the comparative analysis approach for detailed components involves systematic evaluation of specific technical alternatives against precisely defined criteria. This application of techniques from Sect. 4.2.1 includes quantitative performance measures, economic factors, reliability metrics, and compatibility requirements to identify optimal component designs and configurations.

Role-Specific Expertise Through Role Prompting

Applying role prompting (Sect. 4.2.2.3) for detailed design involves directing the LLM to adopt highly specialized technical perspectives relevant to specific components or subsystems. By instructing the model to respond as a heat exchanger design specialist, pump selection expert, or control valve engineer, this technique leverages domain-specific knowledge for optimizing detailed components.

Tree of Thoughts for Parameter Optimization

The Tree of Thoughts technique (Sect. 4.2.3.1) can be applied to parameter optimization by exploring multiple parameters set simultaneously, evaluating their implications, and identifying optimal combinations. This approach is particularly valuable for complex industrial components where multiple parameters interact to determine overall performance.

Chain of Thought for Technical Risk Assessment

Applying Chain of Thought (Sect. 4.2.2.1) to technical risk assessment involves systematic reasoning about potential failure modes, operational limitations, and design uncertainties. By guiding the LLM through structured analysis of technical risks, this approach identifies potential issues before implementation and develops appropriate mitigation strategies.

4.3.5 Implementation Planning Phase: Pathways to Realization

The implementation planning phase in industrial projects involves developing practical pathways to execute the designed solution, including phasing strategies, risk management approaches, resource allocation, and adaptation mechanisms. This phase bridges the gap between technical design and practical execution, addressing the complexities of bringing industrial systems into operational reality.

Tree of Thoughts for Implementation Pathways

Applying the Tree of Thoughts technique (Sect. 4.2.3.1) to implementation planning involves mapping multiple potential implementation sequences, evaluating their implications, and identifying robust pathways. This approach creates a branching structure of implementation options with assessment of each pathway's risks, resource requirements, and timeline implications.

Scenario-Based Applications of Tree of Thoughts

This specialized application of Tree of Thoughts (Sect. 4.2.3.1) tests implementation strategies against different future scenarios. By evaluating how implementation plans perform under various conditions (market changes, technology developments, regulatory evolution), this approach develops robust strategies that can adapt to changing circumstances rather than optimizing for a single projected future.

Chain of Thought for Resource Optimization

Applying Chain of Thought reasoning (Sect. 4.2.2.1) to resource optimization involves systematic analysis of personnel requirements, equipment needs, financial allocations, and scheduling constraints. This structured approach identifies efficient resource allocation strategies that align with implementation priorities while respecting organizational limitations.

4.4 Technique Selection for Industrial Challenges

The diverse array of prompt engineering techniques presents both opportunities and challenges for industrial prompt engineers. While having multiple techniques available enables addressing a wide range of industrial challenges, selecting the most appropriate approach for specific situations can prove daunting. This section provides guidance on technique selection based on problem characteristics, project phase, and desired outcomes.

4.4.1 Matching Techniques to Project Phases

Different project phases present distinct cognitive challenges that are best addressed by specific prompt engineering techniques. Table 4.1 provides a guide on technique selection across project phases, identifying primary and secondary techniques for each phase based on their alignment with cognitive demands.

Table 4.1 Technique selection by project phase

Project phase	Primary techniques	Secondary techniques	Selection rationale
Problem definition	Context-rich framing, multi-perspective analysis, chain of thought	Role prompting, zero-shot CoT	These techniques excel at problem exploration and framing without premature solution focus
Conceptual design	Zero-shot CoT, tree of thoughts, role prompting	Few-shot prompting, self-consistency	These techniques facilitate broad exploration of solution space with appropriate structure
System architecture	Chain of thought, tree of thoughts, self-consistency	Few-shot prompting, role prompting	These techniques support systematic development of system structure with appropriate integration
Detailed design	Comparative analysis, role prompting, tree of thoughts	Chain of thought, few-shot prompting	These techniques enable precise specification while ensuring technical rigor
Implementation planning	Tree of thoughts, chain of thought, self-consistency	Role prompting, few-shot prompting	These techniques manage complexity of execution planning while ensuring robustness

Effective technique selection considers both the primary cognitive objectives of each phase and the relationships between them. Techniques used in earlier phases should create appropriate foundations for subsequent one, ensuring coherent progression throughout the project lifecycle.

4.4.2 Decision Framework for Technique Selection

Beyond project phases, technique selection should consider specific problem characteristics that influence technique effectiveness. Table 4.2 provides a decision framework based on key problem attributes, guiding selection based on problem complexity, domain specificity, time horizon, and other factors.

This framework provides starting guidance, but experience and practice should adapt technique selection based on specific organizational contexts, technical domains, and experience with different approaches. Maintaining a library of successful prompts with associated outcomes can further refine technique selection over time.

Table 4.2 Technique selection based on problem characteristics

Problem characteristic	Recommended techniques	Techniques to avoid
High complexity with multiple variables	Tree of thoughts, self-consistency, multi-perspective analysis	Simple prompting, one-shot examples
Well-defined technical parameters	Chain of thought, comparative analysis	Open-ended exploration techniques
Ambiguous requirements	Multi-perspective analysis, context-rich framing	Comparative analysis, simple prompting
Long-term strategic horizon	Tree of thoughts, scenario-based applications	Simple prompting
Safety–critical applications	Self-consistency, chain of thought	Simple prompting
Cross-disciplinary integration	Multi-perspective analysis, role prompting (multiple roles)	Role prompting (single perspective)
Novel technical domains	Zero-shot CoT, role prompting, tree of thoughts	Simple prompting, one-shot examples
Resource-constrained implementation	Tree of thoughts with phased approach, chain of thought	Simple prompting

4.4.3 Combining Techniques for Complex Problems

The most sophisticated prompt engineering applications frequently combine multiple techniques to address different aspects of complex industrial challenges. Effective combinations leverage complementary strengths while mitigating individual limitations. Table 4.3 presents successful technique combinations for common industrial challenges.

When combining techniques, the following factors should be considered:

1. Logical sequencing of techniques within the prompt
2. Clear transitions between different analytical approaches
3. Appropriate depth allocation to each technique component
4. Consistent terminology and frameworks across the prompt
5. Clear guidance on how to integrate insights from different techniques.

Table 4.3 Effective technique combinations for industrial challenges

Industrial challenge	Recommended technique combination	Integration approach
Process optimization with multiple objectives	Role prompting + Tree of thoughts + Chain of thought	Use role prompting to establish expert perspective, tree of thoughts to explore trade-offs, and chain of thought for detailed analysis
Safety analysis for complex systems	Chain of thought + Self-consistency + Multi-perspective analysis	Apply chain of thought for systematic hazard identification, self-consistency for verification, and multi-perspective analysis for comprehensive coverage
Technology selection for major capital projects	Context-rich framing + Tree of thoughts + Comparative analysis	Begin with context framing to establish evaluation criteria, use tree of thoughts to explore technology options, and apply comparative analysis for final selection
Environmental compliance strategy	Multi-perspective analysis + Tree of thoughts + Chain of thought	Use multi-perspective analysis to understand stakeholder requirements, tree of thoughts to explore regulatory scenarios, and chain of thought for detailed compliance planning
Digital transformation roadmap	Context-rich framing + Tree of thoughts + Chain of thought	Start with comprehensive context, use tree of thoughts for technology options, chain of thought for implementation planning

Thoughtful technique combination can create capabilities significantly beyond what any individual technique might achieve, particularly for the multifaceted challenges typical of industrial transformation. Appendix B provides a complete set of prompt engineering examples applied to many industrial processes and disciplines illustrating all the information explained in the previous sections about prompt engineering techniques and strategies.

4.5 Preventing Hallucinations in Technical Applications

4.5.1 Understanding Hallucination Mechanisms in Technical Domains

Hallucination is the phenomenon where Large Language Models generate plausible but factually incorrect information and poses a particular challenge for industrial applications in petroleum refining and related technical fields. When using LLMs for complex engineering tasks, inaccurate outputs could lead to flawed designs, inefficient processes, or even safety hazards. Understanding the mechanics of these hallucinations is essential for developing effective mitigation strategies.

Hallucinations in LLMs typically stem from several interrelated factors:

1. **Knowledge limitations**: When operating beyond the model's training data or knowledge boundaries, especially for specialized technical information or recent developments
2. **Over generalization**: The model's statistical tendency to fill knowledge gaps with plausible sounding but potentially incorrect information
3. **Contextual misalignment**: Misinterpretation of the task or the significance of certain domain-specific terms
4. **Abstraction challenges**: Difficulties in translating abstract concepts into concrete, technically accurate specifications.

In petroleum refining applications, hallucinations might manifest as incorrect process parameters, unrealistic equipment specifications, improper integration of unit operations, or erroneous thermodynamic relationships. The consequences of such errors could range from inefficient designs to serious safety issues.

4.5.2 Constraint Based Prompting

One of the most effective techniques for mitigating hallucinations in technical contexts is constraint-based prompting. This approach explicitly defines the boundaries and constraints within which the model should operate.

Example 1: Constraint Based Prompting for Process Design
Weak prompt (susceptible to hallucination): "*Design a hydrotreating process for diesel production.*"

Strong constraint-based prompt: "*Design a hydrotreating process for diesel production with the following constraints: Feed properties: 500–5000 ppm sulfur, 20–100 ppm nitrogen, API gravity 30–36; Product specifications: < 10 ppm sulfur, < 5 ppm nitrogen; Operating* *parameters: Temperature range 300–400 °C, pressure range 50–100 bar; Catalyst: NiMo or CoMo supported on alumina; LHSV range: 0.5–2.0 h^{-1}; Hydrogen consumption: 200–400 Nm^3/m^3. For any parameters where precise values are needed but not specified above, explicitly state your assumptions and justify them based on established industry practice. Identify which aspects of your design have high certainty versus which might require further validation.*"

By clearly specifying the technical parameters and their acceptable ranges, the prompt constrains the model's response space to physically and technically plausible solutions, reducing the likelihood of hallucination.

4.5.3 First Principles Reasoning

Instructing LLMs to apply first principles reasoning helps anchor their responses in fundamental scientific and engineering concepts rather than pattern matching from potentially incomplete training data.

Example 2: First Principles Reasoning for Energy Integration

Weak prompt: "*How can we optimize heat integration between a crude distillation unit and a hydrocracker?*"

Strong first principles prompt: "*Using the principles of thermodynamics and heat transfer, systematically analyze the opportunities for heat integration between a crude distillation unit and a hydrocracker. Follow these steps: 1. Identify the key hot streams (requiring cooling) and cold streams (requiring heating) in each unit, 2. For each stream, specify the typical temperature ranges and heat capacities, 3. Apply pinch analysis principles to determine the minimum utility requirements, 4. Propose specific heat exchanger network configurations based on temperature driving forces, 5. Identify technical limitations (e.g., pressure differentials, fouling considerations) that might constrain integration, 6. Quantify potential energy savings using established engineering methods. Base your analysis exclusively on fundamental engineering principles and explicitly note any areas where additional data would be required for detailed design.*"

This approach guides the model to construct its response from basic scientific principles, reducing reliance on potentially incomplete or inaccurate memorized patterns.

4.5.4 Explicit Uncertainty Disclosure

Requiring the model to explicitly acknowledge uncertainty helps distinguish between high confidence knowledge and areas where the model is less certain, providing crucial context for evaluating its outputs.

Example 3: Uncertainty Disclosure for Catalyst Selection

Compare platinum, palladium, and rhenium catalysts for catalytic reforming applications in petroleum refining. For each catalyst: 1. Describe the fundamental catalytic mechanisms, 2. Specify typical operating conditions, 3. Identify strengths and limitations, 4. Estimate typical performance metrics. Clearly indicate your level of confidence in each statement using the following system: [High Confidence]: Statements based on well-established scientific literature; [Medium Confidence]: Statements based on general principles but with potential variations; [Low Confidence]: Areas where more specific information would be needed for a definitive answer. When referencing quantitative parameters, provide ranges rather than precise figures unless the values are firmly established in the literature.

This approach encourages the model to be transparent about its knowledge limitations, helping users distinguish between reliable information and areas requiring verification.

4.5.5 Verification Prompting

Verification prompting involves instructing the model to critically evaluate its own responses, simulating the verification process a human expert might apply.

Example 4: Self-verification for Process Safety Analysis

Conduct a process hazard analysis for the following scenario: hydrogen leakage in a hydrocracker unit. After completing your initial analysis, verify your response by: 1. Assessing each identified hazard against established process safety principles, 2. Checking that all recommended controls follow the hierarchy of controls, 3. Verifying that all calculations (e.g., explosion indices, safe distances) follow correct methodology, 4. Identifying any statements that require additional data or expertise to validate, 5. Highlighting any areas where industry standards or regulatory requirements would need to be consulted. Present both your original analysis and your verification assessment, noting any revisions or qualifications resulting from the verification process.

By requiring the model to apply validation criteria to its own output, verification prompting can identify potential hallucinations before they reach the user.

4.5.6 Knowledge Boundary Establishment

Explicitly defining the knowledge boundaries helps prevent the model from generating speculative content beyond its training data.

Example 5: Knowledge Boundary Establishment for Emerging Technologies

Provide an assessment of current commercial catalytic pyrolysis technologies for plastic waste conversion in petroleum refineries. Limit your response to technologies that were documented in the scientific or trade literature prior to October 2024. For each technology, include: 1. Developer/licensor, 2. Core technological principles, 3. Feedstock requirements, 4. Reported yields and product characteristics, 5. Current deployment status. If certain information is not broadly available in the public domain (e.g., proprietary process details), explicitly note these knowledge gaps rather than speculating. If a technology emerged after your knowledge cutoff, indicate that your information may be incomplete.

This approach establishes clear temporal and subject matter boundaries, reducing the likelihood of the model generating speculative information about recent developments or proprietary technologies.

4.5.7 Comparative Analysis Requirements

Requiring structured comparison across multiple alternatives forces the model to maintain consistency and adhere to physical and technical realities.

Example 6: Comparative Analysis for Desulfurization Technologies

Compare the following three desulfurization technologies for petroleum refining: 1. Conventional hydrodesulfurization, 2. Oxidative desulfurization, 3. Biodesulfurization. For each technology, analyze using identical criteria: Reaction chemistry and mechanisms; Typical operating conditions (temperature, pressure); Catalyst or biological agent requirements; Energy intensity (in GJ/ton of treated feedstock); Sulfur removal efficiency for different sulfur compounds; Capital and operating cost considerations; Commercial implementation status. Present your analysis in a structured table followed by comparative discussion. Where quantitative data varies across different implementations, provide representative ranges with sources of variation.

By requiring parallel analysis across multiple technologies, the prompt creates internal consistency checks that help identify and mitigate potential hallucinations.

4.5.8 Application in Petroleum Refining

The techniques described above are particularly valuable for petroleum refining applications, where technical precision is critical. When applying these techniques to refinery operations, process engineering, or sustainability transformation, practitioners should:

1. **Prioritize domain specificity**: Include industry standard terminology, units, and reference points to ground the model in the petroleum refining context

2. **Incorporate relevant standards**: Reference applicable industry standards (API, ASTM, etc.) to guide the model toward conformance with established practices
3. **Layer multiple techniques**: Combine approaches such as constraint-based prompting with verification requirements for maximum effectiveness
4. **Validate critical outputs**: Use conventional engineering tools to verify key calculations or relationships generated by the model, particularly for safety–critical applications.

The systematic application of these hallucination mitigation techniques transforms LLMs from interesting but potentially unreliable tools into valuable engineering assistants capable of supporting complex technical tasks while maintaining appropriate levels of accuracy and reliability.

4.6 The Future Evolution of Prompt Engineering in Industry

The evolutionary development of prompt engineering techniques continues, with emerging approaches addressing current limitations while creating new capabilities. Understanding these evolving approaches provides insight into potential future developments in this rapidly advancing field.

4.6.1 From Prompt Engineering to AI Augmented Design

The long-term trajectory of prompt engineering points toward a fundamental shift from isolated prompting techniques to comprehensive AI-augmented design methodologies, in this vein we hace to mention the concept of Context Engineering introduced by Andej Karpathy as "The delicate art and science of filling the context window with just the right information for the next step", this is the natural evolution of prompt engineering and has been developed at the same time this book was written; fortunately many of the concepts introduced in this chapter about the relevance of the context to improve the prompt performace anticipated the Context Engineering concepts and practices. This evolution will likely integrate LLMs as central components within broader design systems that include:

1. Specialized simulation and calculation tools
2. Knowledge management systems
3. Collaborative workflows across human-AI teams
4. Continuous learning from implementation outcomes
5. Integration with physical systems and sensors.

This integration will transform prompt engineering from a discrete skill to a component of comprehensive AI-augmented design practices that span the entire industrial innovation cycle from concept development through detailed design to implementation and operation.

The implications for industrial transformation are profound. As these capabilities mature, they will enable:

1. Dramatic acceleration of design cycles through rapid exploration
2. More comprehensive assessment of alternatives before commitment
3. Enhanced integration across disciplinary boundaries
4. More systematic capture and application of organizational knowledge
5. Continuous improvement based on operational feedback.

While this evolution may significantly enhance engineering capabilities, it should be viewed as a complementary approach rather than a paradigm replacement. Industrial transformation still fundamentally relies on physical testing, regulatory compliance, safety verification, and field experience that cannot be virtualized or accelerated through AI systems alone. The integration of AI tools presents promising opportunities for improvement, but successful implementations will depend equally on organizational adaptation, skill development, and careful validation processes.

Figure 4.9 illustrates the progression from isolated prompt engineering activities to a fully integrated AI-augmented design ecosystem, showing how LLMs become central components in a network of specialized tools, knowledge bases, human collaborators, and physical systems. The visualization captures how the role of prompt engineering expands from advisory interactions to leading complex workflows across the entire industrial innovation lifecycle always under human supervision.

4.6.2 Refinery-Specific Future Applications

The future evolution of prompt engineering has particularly promising applications for petroleum refining transformation:

Process Simulators and Digital Twin Integration: Advanced prompt engineering techniques will increasingly interface with refinery process simulators and digital twins, enabling natural language interactions with simulation models. Engineers will query complex operational scenarios through conversational interfaces rather than manual simulation configuration, dramatically accelerating scenario testing and optimization.

Autonomous Process Optimization: The integration of LLMs with advanced control systems will create semi-autonomous optimization capabilities that continuously adapt

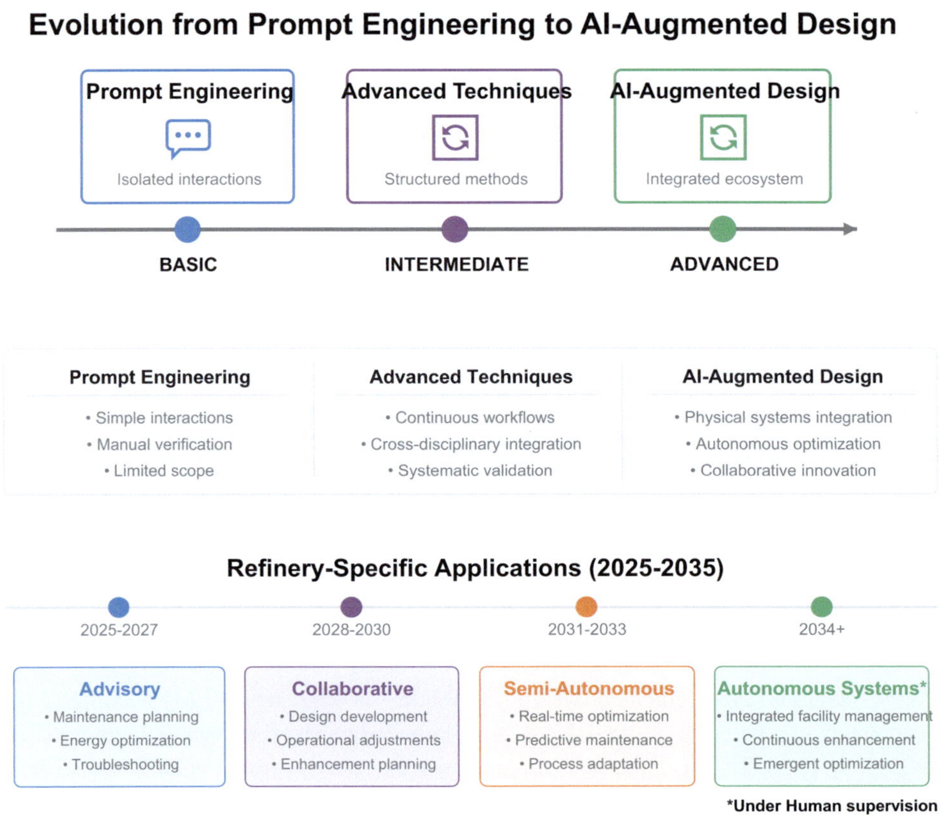

Fig. 4.9 Evolution from prompt engineering to AI-augmented design ecosystems

to changing conditions. These systems will leverage historical performance data, current market conditions, and predictive models to suggest optimal operating parameters in real-time.

Knowledge Preservation and Transfer: As the refining industry faces challenges with experienced personnel retiring, prompt engineering will play a crucial role in knowledge management. Sophisticated RAG systems will capture and operationalize decades of knowledge, making it accessible to new generations of engineers through natural language interfaces.

4.7 Conclusion

The systematic framework for industrial prompt engineering presented in this chapter provides a comprehensive approach to leveraging these capabilities throughout the project lifecycle, from problem definition through conceptual design, system architecture, detailed specification, and implementation planning.

The development of prompt engineering techniques from simple instructions through reasoning focused approaches to advanced problem exploration methods provides a toolkit for addressing diverse industrial challenges. By systematically mapping these techniques to specific project phases and problem characteristics, organizations can develop increasingly sophisticated prompt engineering capabilities that deliver consistent value across applications.

Adopting the structured methodology presented in this chapter, organizations can transform LLMs into key assets for industrial applications. This change requires not just technical understanding of prompt engineering techniques but also appropriate implementation approaches addressing context management, knowledge transfer, quality assurance, and organizational integration. Organizations that begin developing these capabilities today will be best positioned to leverage this evolution, accelerating their innovation processes while enhancing their ability to address complex transformation challenges.

References

Brown TB, Mann B, Ryder N, Subbiah M, Kaplan J, Dhariwal P, Neelakantan A, Shyam P, Sastry G, Askell A et al (2020) Language models are few-shot learners. Adv Neural Inf Process Syst 33:1877–1901

Chowdhery A, Narang S, Devlin J, Bosma M, Mishra G, Roberts A, Barham P, Chung HW, Sutton C, Gehrmann S, Schuh P (2022) PaLM: scaling language modeling with pathways. arXiv preprint arXiv:2204.02311

Clark K, Khandelwal U, Levy O, Manning CD (2019) What does BERT look at? An analysis of BERT's attention. arXiv preprint arXiv:1906.04341

Kojima T, Gu SS, Reid M, Matsuo Y, Iwasawa Y (2022) Large language models are zero-shot reasoners. arXiv preprint arXiv:2205.11916

Lewis P, Perez E, Piktus A, Petroni F, Karpukhin V, Goyal N, Küttler H, Lewis M, Yih W, Rocktäschel T, Riedel S (2020) Retrieval-augmented generation for knowledge-intensive NLP tasks. arXiv preprint arXiv:2005.11401

Liu P, Yuan W, Fu J, Jiang Z, Hayashi H, Neubig G (2021) Pre-train, prompt, and predict: a systematic survey of prompting methods in natural language processing. ACM Comput Surv 55(9):1–35

OpenAI (2023) GPT-4 technical report. arXiv preprint arXiv:2303.08774

Ouyang L, Wu J, Jiang X, Almeida D, Wainwright C, Mishkin P, Lowe R (2022) Training language models to follow instructions with human feedback. Adv Neural Inf Process Syst 35:27730–27744

Ram O, Levine Y, Dalmedigos I, Muhlgay D, Shashua A, Leyton-Brown K, Shoham Y (2023) In Context Retrieval Augmented Language Models. Trans Assoc Comput Lingu 11:1316–1331

Reynolds L, McDonell K (2021) Prompt programming for large language models: beyond the few-shot paradigm. arXiv preprint arXiv:2102.07350

Vaswani A, Shazeer N, Parmar N, Uszkoreit J, Jones L, Gomez AN, Kaiser Ł, Polosukhin I (2017) Attention is all you need. Adv Neural Inf Process Syst 30, 6000-6010

Wang X, Wei J, Schuurmans D, Le Q, Chi E, Zhou D (2022) Self-consistency improves chain of thought reasoning in large language models. arXiv preprint arXiv:2203.11171

Wei J, Wang X, Schuurmans D, Bosma M, Ichter B, Xia F, Zhou D (2022) Chain-of-thought prompting elicits reasoning in large language models. Adv Neural Inf Process Syst 35:24824–24837

White J, Fu Q, Hays S, Sandborn M, Olea C, Gilbert H, Elnashar A, Spencer-Smith J, Schmidt DC (2023) A prompt pattern catalog to enhance prompt engineering with ChatGPT. arXiv preprint arXiv:2302.11382

Yao S, Yu D, Zhao J, Shafran I, Griffith TL, Cao Y, Narasimhan K (2023) Tree of thoughts: deliberate problem solving with large language models. arXiv preprint arXiv:2305.10601

Zamfirescu-Pereira JD, Wong RY, Hartmann B, Yang Q (2023) Why Johnny can´t prompt: Why Non -AI Experts Try and Fail to Design LLM Prompts. Proceedings of the 23 CHI Conference of Human Factors in Computing Systems, Article 437, 1–21

Zhou D, Schärli N, Hou L, Wei J, Scales N, Wang X, Schuurmans D, Bousquet O, Le Q, Chi E (2022) Least-to-most prompting enables complex reasoning in large language models. arXiv preprint arXiv:2205.10625

5 Prompting Net Zero Refinery Design

Abstract

This chapter demonstrates the practical application of prompt engineering methodologies to design a comprehensive net zero refinery. The design process follows industrial project phases from problem definition through conceptual design, system architecture, detailed design, and implementation planning. Using systematic prompt engineering techniques including context-rich framing, multi perspective analysis, tree of thoughts, and chain of thought reasoning, a Large Language Model helps to develop a viable refinery concept that processes diverse alternative feedstocks through integrated conversion technologies powered by renewable energy. The resulting design demonstrates potential for carbon negative performance while addressing the technical, economic, and implementation challenges inherent in such a transformation. This chapter provides concrete examples of how Large Language Models can contribute to complex industrial innovation when properly supervised and directed through structured prompting methodologies, with particular attention to realistic system sizing and practical implementation considerations.

5.1 Introduction: The Challenge of Net Zero Refining

This chapter demonstrates the practical application of prompt engineering methodologies established in Chap. 4 to design a comprehensive net zero refinery. Unlike theoretical discussions, we present the actual prompts used with Large Language Models (LLMs) and study their resulting outputs, illustrating how systematic prompt engineering under human

Rafael Larraz, Prompt Engineering and the Transformation of Petroleum Refining.

R. Larraz, *Prompt Engineering and the Transformation of Petroleum Refining*, Synthesis Lectures on Chemical Engineering and Biochemical Engineering,
https://doi.org/10.1007/978-3-031-99728-0_5

supervision can accelerate complex industrial innovation while addressing sustainability challenges.

The design process follows the industrial project phases outlined in Chap. 4: from problem definition through conceptual design, system architecture, detailed design, and implementation planning. At each stage, appropriate prompt engineering techniques is applied to address the specific cognitive demands of that phase, while carefully evaluating and refining the LLM outputs based on engineering judgment and industrial experience.

This progressive development enabled a viable refinery concept that balances technological opportunities with practical constraints. Importantly, the process revealed how LLM outputs presents errors and require critical evaluation and iterative refinement based on engineering analysis to achieve realistic proportions.

5.2 Methodology Application Framework

Before diving into the design process, it's essential to establish how specific prompt engineering techniques from Chap. 4 are applied across the design phases. Figure 5.1 illustrates this application framework, where each design phase employs specific prompt engineering techniques most suited to its cognitive demands.

The framework ensures that prompt engineering techniques are selected based on the analytical requirements of each phase rather than arbitrary application. This systematic approach proved essential for maintaining design coherence while leveraging the distinct strengths of different prompting methodologies. In the following sections we present the prompts employed and an analysis of the results obtained oriented to understand the LLM performance. The complete set of results can be found in Appendix C.

5.3 Problem Definition: Establishing Design Boundaries

The problem definition phase establishes the foundation for subsequent design activities by clearly defining system boundaries, identifying constraints, analyzing stakeholder perspectives, and establishing performance objectives. This phase proved crucial for ensuring that the design addresses the appropriate problem scope and incorporates all relevant contextual factors before beginning solution development.

5.3.1 Context-Rich Framing

To establish comprehensive system boundaries, we employed the context-rich framing technique, which provides detailed background information enabling the LLM to develop

Comprehensive Industrial Prompt Engineering Framework

Project Phase	Primary Objectives	Key Prompting Techniques	Expected LLM Contributions
Problem Definition	System boundary definition Constraint mapping Stakeholder analysis	Context-rich framing Multi-perspective analysis Constraint elicitation	Comprehensive problem statements Complete constraint identification
Conceptual Design	Design space exploration Solution principles Technology scanning	Zero-shot chain of thought Few-shot ideation Tree of thoughts	Alternative conceptual approaches Cross-domain solution principles
System Architecture	Functional decomposition Process flow development Integration challenge	Chain of thought Comparative analysis Self-consistency	System structure alternatives Function allocation schemes Interface requirements
Detailed Design	Component specification Performance optimization Technical risk assessment	Comparative analysis Role-specific expertise Quantitative parameters	Detailed design parameters Performance-constraint analysis
Implementation Planning	Pathway development Risk management Resource optimization	Tree of thoughts Scenario-based exploration Decision criteria enumeration	Phased implementation strategies Risk management approaches

Fig. 5.1 Industrial prompt engineering framework showing the systematic application of prompt engineering techniques across design phases. Each phase employs specific techniques suited to its cognitive demands, ensuring optimal matching of methodology to requirements

responses grounded in specific circumstances rather than generic assumptions. This technique proved particularly valuable for avoiding overly theoretical responses that might not reflect real-world constraints.

Prompt Applied:

We aim to design a net zero refinery to replace an existing 100,000 barrels per day petroleum refinery located in a coastal setting surrounded by urban areas, agricultural zones, and industrial facilities. The region has established waste collection systems processing approximately 1 million tons annually of municipal solid waste, 25% of which is currently recycled. Agricultural operations within 200 km produce approximately 2 million tons of crop residues annually, most currently left in fields or burned. The electrical grid has substantial capacity (200 MW connection possible) with a moderately decarbonized generation mix (350 g CO_2/kWh).

Given this context, define the system boundaries for a net zero refinery that must produce transportation fuels (sustainable aviation fuel, renewable diesel, and renewable gasoline) and chemical intermediates without petroleum inputs while achieving carbon neutrality or negativity on a lifecycle basis. Identify all primary material, energy, and information flows that will cross this system boundary.

LLM Output Analysis: The LLM response provided a comprehensive system boundary definition that incorporated the specific contextual factors provided. The output identified three core principles for the system design:

1. **Circular Resource Utilization**: Converting locally available waste streams into valuable products
2. **Integrated Carbon Management**: Strategic deployment of carbon capture and utilization
3. **Regional Industrial Symbiosis**: Leveraging existing infrastructure.

Critical Evaluation and Refinement: The LLM output was generally sound but required human intervention and refinement in several areas. The initial response underestimated the complexity of feedstock preprocessing and overestimated the ease of carbon management integration. Based on industrial experience, we refined the system boundaries to explicitly include:

- Buffer storage systems to manage feedstock seasonality
- More detailed utility systems for the diverse processing requirements
- Enhanced waste heat recovery networks
- Realistic carbon accounting systems with third party verification.

This comprehensive system boundary definition established the framework for the net zero refinery, emphasizing that achieving true net zero requires integrating multiple carbon management strategies beyond just replacing fossil feedstocks.

5.3.2 Multi-perspective Analysis

To ensure comprehensive understanding of requirements and constraints from diverse viewpoints, we employed multi-perspective analysis, explicitly requesting consideration from multiple technical, operational, and stakeholder perspectives. This technique proved essential for identifying potential conflicts and constraints that might not be apparent from a single viewpoint.

Prompt Applied:

Analyze the net zero refinery design challenge from the following stakeholder perspectives:

1. *Technical perspective: Core technologies required, integration challenges, performance metrics*
2. *Economic perspective: Capital requirements, operating economics, market considerations*
3. *Environmental perspective: Carbon lifecycle, local impacts, resource utilization*
4. *Community perspective: Employment impacts, sensory impacts, regional development*
5. *Regulatory perspective: Compliance pathways, permitting requirements, policy engagement*
6. *Operational perspective: Reliability requirements, maintenance considerations, safety factors.*

   For each perspective, identify:

(a) *Primary requirements that must be satisfied*
(b) *Key constraints that limit design options*
(c) *Critical success factors for implementation*
(d) *Potential conflicts with other perspectives*
(e) *Metrics for evaluating success from this perspective.*

Conclude with an integrated framework that balances these perspectives.

LLM Output Analysis: The LLM provided a structured analysis from each perspective, identifying distinct priorities and potential conflicts. Particularly valuable was the identification of tensions between economic optimization (favoring proven technologies) and environmental goals (requiring advanced, less mature technologies).

Critical Evaluation and Decision Making: The multi-perspective analysis revealed several critical insights that shaped subsequent design decisions:

1. **Technology Maturity Conflicts**: Environmental goals pushed toward cutting-edge technologies while economic and operational perspectives favored proven systems. This led to the decision to implement a phased approach with mature technologies in early phases.
2. **Community Acceptance Requirements**: The community perspective emphasized job creation and sensory impact management, leading to specific design requirements for noise control and local employment programs.
3. **Regulatory Complexity**: The analysis revealed the need for early regulatory engagement across multiple jurisdictions and regulatory frameworks.

The results of this analysis are summarized in Table 5.1, which presents key insights from each stakeholder perspective.

Table 5.1 Multi-perspective analysis summary

Perspective	Primary requirements	Key constraints	Critical success factors	Key metrics
Technical	Core technologies (gasification, pyrolysis, Fischer–Tropsch), seamless integration, > 90% uptime	Feedstock variability, technology maturity, grid limitations	Robust preprocessing, advanced process control, pilot testing	Conversion efficiency (> 80%), system uptime (> 90%), CO_2 capture rate (> 90%)
Economic	$2–3 billion capital investment, competitive production costs, market viability	High CapEx for advanced technologies, feedstock costs, market risks	Government incentives, economies of scale, diversified revenue	Return on Investment (> 10%), levelized cost (< $1.5/l), revenue stability (> 80% secured by contracts)
Environmental	Net zero/negative emissions, minimal local pollution, resource efficiency	Grid carbon intensity, feedstock competition, CO_2 storage limitations	CCUS deployment, renewable energy integration, third-party verification	Lifecycle GHG emissions (≤ 0 g CO_2e/MJ), local emissions (< 100 tons/year NOx/SOx)
Community	Job creation (500–1000 direct jobs), minimal sensory impacts, regional development	Urban proximity, public perception concerns, budget limitations	Transparent engagement, local hiring, mitigation measures	Job creation (>500 direct jobs), community complaints (< 10/ year)
Regulatory	Compliance with regulations, necessary permits, policy alignment	Complex permitting processes, evolving policies, strict local regulations	Early engagement with regulators, robust documentation, policy advocacy	Permitting timeline (< 18 months), 100% compliance, incentive capture (> $50 M/ year)
Operational	> 90% uptime, manageable maintenance, safety excellence	Technology complexity, feedstock variability, workforce skills	Digital tools, training programs, safety culture	Uptime (> 90%), maintenance cost (< $2/bbl), zero process safety incidents

The integrated framework that emerged from this analysis adopted a phased approach that prioritizes early regulatory and community engagement, leverages proven technologies with modular scaling, and optimizes operations for reliability and profitability.

5.3.3 Constraint Mapping

To establish clear design boundaries, a chain of thought reasoning was used to systematically identify constraints and limitations across multiple dimensions that would shape the design solution space. This systematic approach proved essential for avoiding designs that might be technically interesting but practically impossible.

Prompt Applied:

Conduct a systematic constraint mapping for a net zero refinery design across the following dimensions:

1. Physical Constraints:
   - *Analyze available land within refinery "red line" boundaries*
   - *Identify infrastructure connection limitations*
   - *Determine geographical and location restrictions*
   - *Consider maximum processing system capacities*
   - *Map proximity requirements for critical unit operations.*

[Additional constraint dimensions covered feedstock, technical, energy, and regulatory constraints with similar systematic analysis requested].
For each constraint category, identify:

(a) *Hard constraints that cannot be compromised*
(b) *Soft constraints that permit design trade-offs*
(c) *Critical dependencies between constraint categories*
(d) *Potential mitigation approaches for key limitations.*

Present your analysis as a structured constraint map with quantitative parameters where possible and qualitative assessments where necessary.
LLM Output Analysis: The LLM provided a comprehensive constraint analysis that identified both hard and soft constraints across all dimensions. Particularly valuable was the identification of interdependencies between constraint categories.

Critical Evaluation and Refinement: The constraint analysis required significant refinement based on industrial experience. The LLM initially underestimated several practical constraints:

1. **Feedstock Logistics**: The LLM initially assumed uniform feedstock availability, but analysis revealed significant seasonal variations requiring 6–9 months of buffer storage.
2. **Technology Integration**: The complexity of integrating diverse conversion technologies was initially underestimated, leading to revised requirements for intermediate storage and processing flexibility.
3. **Energy System Constraints**: The scale requirements for renewable energy integration proved larger than initially estimated, requiring careful consideration of grid interconnection and land use.

The systematic constraint analysis revealed several critical limitations that shaped the final design:

- **Physical Constraints**: Available land within refinery boundaries, infrastructure connection limitations
- **Feedstock Constraints**: Seasonal availability requiring substantial buffer storage, moisture content limitations, economic transport radius of 150 km maximum
- **Technical Constraints**: Technology readiness levels varying from TRL 5–9, scaling limitations, integration complexity
- **Energy Constraints**: Electrification vs. grid reliability trade-offs, renewable intermittency management

Critical trade-offs identified included electrification versus grid reliability (full electrification requires redundant power infrastructure) and land vs. preprocessing (compact refineries force offsite biomass preprocessing, adding 15–20% logistics costs).

5.3.4 Problem Definition Synthesis

The problem definition phase established a comprehensive foundation for the net zero refinery design through systematic application of prompt engineering techniques. The process demonstrated both the value and limitations of LLM-assisted design:

Strengths Observed:

- Comprehensive perspective integration across stakeholder groups
- Systematic identification of constraints and interdependencies
- Structured approach to complex, multi-dimensional problems.

Limitations Requiring Human Judgment:

- Initial outputs often lacked practical scaling considerations
- Technology integration complexity was consistently underestimated
- Economic realities required significant refinement of initial concepts.

Key insights that emerged from this phase include:

1. **System Integration Imperative**: The net zero refinery must function as an integrated system crossing traditional boundaries between energy, waste management, and industrial processing.
2. **Multi-stakeholder Complexity**: Successful implementation requires balancing diverse and sometimes conflicting stakeholder requirements, with economic viability necessary but insufficient without addressing community, regulatory, and operational concerns.
3. **Constraint Hierarchies**: Critical hard constraints (such as land availability and feedstock characteristics) cannot be compromised, while soft constraints permit design trade-offs.

These insights provided a robust foundation for the conceptual design phase, ensuring that solution exploration would be grounded in a comprehensive understanding of the problem space.

5.4 Conceptual Design: Exploring Technological Pathways

The conceptual design phase explores the solution space to identify promising approaches before detailed development begins. This phase involves broad exploration of potential solution principles, technologies, and architectural approaches without premature commitment to specific implementations.

5.4.1 Feedstock Conversion Pathway Analysis

To systematically explore potential processing pathways for municipal solid waste, we selected zero-shot chain of thought reasoning, which guides the LLM through structured analysis without requiring specific examples. This technique proved particularly effective for exploratory analysis where the goal was to identify possibilities rather than optimize known solutions.

Prompt Applied:

Using a step-by-step reasoning process, analyze the potential processing pathways for municipal solid waste (MSW) in a net zero refinery. Begin by characterizing the typical composition of municipal solid waste and identify the carbon-containing fractions that could serve as refinery feedstock. Then, systematically evaluate possible conversion pathways for each fraction, considering:

1. *Initial preprocessing requirements*
2. *Primary conversion processes*
3. *Upgrading and finishing steps*
4. *Product yield expectations*
5. *Carbon efficiency of the pathway*
6. *Technology readiness level*
7. *Economic considerations.*

After analyzing individual options, identify the most promising pathways for municipal solid waste conversion and explain your reasoning.

LLM Output Analysis: The LLM provided a systematic breakdown of MSW composition and processing pathways. The response identified several conversion routes including anaerobic digestion, gasification, and pyrolysis, with detailed analysis of each pathway's characteristics.

Critical Evaluation and Decision Making: The LLM output required significant analysis and refinement based on industrial experience and recent literature. The systematic analysis revealed that municipal solid waste represents a highly valuable but complex feedstock containing 341,850 tons of carbon annually across diverse material streams requiring differentiated processing approaches.

Key analytical insights that emerged from evaluating the LLM output:

1. **Feedstock Diversity Challenge**: The 90% carbon-containing fraction spans materials with vastly different properties, from 60% moisture food waste to dry plastics with 80% carbon content, requiring multiple processing pathways rather than a single conversion route.
2. **Technology Readiness Spectrum**: Anaerobic digestion (TRL 9) provides low-risk implementation, while gasification (TRL 6–7) and pyrolysis (TRL 7–8) offer higher yields but increased technical risk, creating a natural phasing strategy.
3. **Techno-Economic Trade-off**: Higher carbon efficiency pathways (pyrolysis at 85%, gasification at 80%) require significantly higher capital investment ($400–650 M) compared to proven anaerobic digestion ($60–90 M), but generate premium products worth $400–600/ton versus $50–80/ton for biogas.

Recommended Strategy Development: Based on the analysis, we developed an integrated anaerobic digestion (AD)+Gasification approach that maximizes both risk management and carbon utilization by:

- Processing 250,000 tons of high-moisture organics through proven AD technology for immediate revenue
- Converting 400,000 tons of dry materials through gasification for high-value synthetic fuels
- Achieving 72–78% overall carbon efficiency while maintaining manageable technical risk
- Generating 15,000–18,000 bbl/day equivalent renewable fuels with reasonable ROI.

This phased approach transforms MSW from a waste disposal problem into a strategic feedstock contributing 30–40% of the refinery's total renewable fuel production while demonstrating the circular economy principles essential for net zero operations.

5.4.2 Technology Pathway Exploration

To explore different decision branches for the net zero refinery design, a tree of thoughts prompting was used, which enables exploration of multiple conceptual branches simultaneously. This technique proved particularly valuable for complex decisions where multiple viable paths exist with different implications.

Prompt Applied:

Explore the decision landscape for net zero refinery design using a tree of thoughts approach.
Start with five major decision branches:
Branch 1: Primary conversion approach

- *Option 1A: Biochemical dominant (fermentation, digestion)*
- *Option 1B: Thermochemical dominant (gasification, pyrolysis)*
- *Option 1C: Hybrid balanced approach.*

Branch 2: Feedstock flexibility

- *Option 2A: Specialized for specific feedstock types*
- *Option 2B: Moderate flexibility with preprocessing*
- *Option 2C: Maximum flexibility for diverse inputs.*

[Additional branches covering product slate focus, energy system design, and carbon management strategy]
For each branch:

1. *Explore the implications of each option*
2. *Identify interdependencies with other branches*
3. *Evaluate how different combinations perform against objectives*
4. *Note particularly promising or problematic combinations.*

After exploring all branches, identify 2–3 most promising overall pathways that represent coherent combinations across these decision branches.
LLM Output Analysis: The LLM provided a comprehensive exploration of each decision branch with analysis of trade-offs and interdependencies. The tree of thoughts approach successfully mapped the complexity of the decision space and identified promising combinations.

Critical Evaluation and Synthesis: The tree of thoughts analysis provided valuable insights into the technological pathway selection but required careful evaluation to distinguish between theoretically possible and practically implementable combinations.

Key insights from the primary conversion approaches:

Option 1A: Biochemical Dominant

- Strengths: Strong performance with organic waste, low-cost biodegradable feedstocks, high potential for low carbon emissions
- Limitations: Limited to organic feedstocks, slower to scale than thermochemical processes
- Assessment: Suitable for organic fraction but insufficient alone.

Option 1B: Thermochemical Dominant

- Strengths: Works with broader feedstock range including non-biological waste and plastics, high scalability and energy efficiency
- Limitations: Balancing high-temperature operations with low-carbon emissions presents challenges
- Assessment: High potential but requires careful energy integration.

Option 1C: Hybrid Balanced Approach

- Strengths: Leverages strengths of both approaches, enables adaptation to wide feedstock range

- Limitations: Requires careful balancing of input types and potentially complex integration
- Assessment: Most promising for comprehensive feedstock utilization.

Final Pathway Selection: The most promising pathway identified was a hybrid approach that combines waste and biomass gasification, plastic pyrolysis, and Fischer–Tropsch synthesis, with integrated carbon capture and renewable energy. This approach balances the advantages of diverse conversion technologies while mitigating their individual limitations through system integration.

5.4.3 Integration Opportunity Analysis

To elicit specialized expertise on process integration opportunities, we employed role prompting, which instructs the LLM to adopt a specific professional perspective. This technique proved particularly effective for accessing domain-specific knowledge and analytical patterns.

Prompt Applied:

Assume the role of a process integration specialist with 25 years of experience in biorefining and petrochemical facilities. You have expertise in heat integration, mass integration, and process intensification across diverse conversion technologies.
Given your expertise, analyze the integration opportunities for the following proposed net zero refinery process blocks:

- *Municipal solid waste sorting and preparation*
- *Plastics catalytic pyrolysis (450–550 °C)*
- *Biomass gasification (850–900 °C)*
- *Syngas cleaning and conditioning*
- *Fischer–Tropsch synthesis (exothermic, 220–250 °C)*
- *Hydroprocessing operations (330–380 °C)*
- *CO_2 capture systems*
- *Hydrogen production.*

For each pair of process blocks that offer significant integration potential:

1. *Identify specific heat integration opportunities*
2. *Analyze mass integration possibilities*
3. *Recommend process intensification approaches*
4. *Quantify potential efficiency improvements from integration*
5. *Note technical challenges or limitations to successful integration*

6. *Suggest practical implementation approaches considering physical distances.*

Provide specific technical recommendations based on your expertise rather than general observations.

LLM Output Analysis: The role prompting successfully activated domain specific knowledge, providing detailed technical recommendations with specific temperature ranges, heat duties, and integration approaches. The response demonstrated understanding of practical implementation challenges.

Critical Evaluation and Refinement: While the LLM output provided valuable integration insights, it required refinement based on practical engineering experience, particularly regarding:

1. **Temperature Matching**: Some proposed heat integration opportunities had insufficient temperature driving forces for practical implementation
2. **Distance Limitations**: Initial recommendations underestimated the impact of physical distances on heat integration effectiveness
3. **Control Complexity**: The integration of multiple process streams requires sophisticated control systems not fully addressed.

Key Integration Opportunities Identified:

1. **Heat Integration Opportunities**:
   - Utilizing waste heat from Fischer–Tropsch synthesis (220–250 °C) to preheat MSW feed and provide auxiliary heat for gasifiers
   - Directing hot gases from pyrolysis (450–550 °C) to preheat feedstock for biomass gasification
   - Using excess heat from Fischer–Tropsch to supply endothermic heat for biomass gasification
   - Implementing closed-loop heat integration for CO_2 capture using waste heat from syngas cleaning.
2. **Mass Integration Opportunities**:
   - Integrating pyrolysis char with CO_2 capture systems to enhance carbon sequestration
   - Directing syngas from gasification into Fischer–Tropsch synthesis to reduce external hydrogen needs
   - Developing modular hydrogen units to directly integrate with hydroprocessing and Fischer–Tropsch.
3. **Process Intensification Recommendations**:
   - Exploring catalytic upgrading of syngas directly in the gasifier
   - Investigating syngas recycling options to improve overall efficiency

- Implementing reactive distillation where applicable.

The integration analysis demonstrated that substantial efficiency improvements (15–25%) are possible through systematic heat, mass, and process integration, highlighting that these integration opportunities are critical design considerations rather than secondary optimizations.

5.4.4 Conceptual Design Synthesis

The conceptual design phase yielded several critical insights that shaped the overall design direction through systematic application of prompt engineering techniques:

Technology Pathway Selection: No single technology offers optimal performance across all metrics, leading to identification of a hybrid approach combining multiple pathways as most promising. This hybrid approach provides flexibility to handle diverse feedstocks while maximizing carbon efficiency and economic viability.

Integration Imperative: Substantial efficiency improvements are possible through systematic heat, mass, and process integration. The analysis revealed opportunities to reduce external energy requirements by 15–25% through strategic integration of exothermic and endothermic processes.

Technology Readiness Considerations: Significant variations in technology readiness levels across different pathways suggest a phased implementation approach that begins with higher-TRL technologies while incorporating more advanced approaches as they mature.

Prompt Engineering Effectiveness Assessment: The conceptual design phase demonstrated both strengths and limitations of prompt engineering approaches:

Strengths:

- Zero-shot chain of thought effectively structured complex analysis
- Tree of thoughts successfully mapped decision complexity
- Role prompting activated relevant domain expertise.

Limitations:

- Initial outputs required significant technical refinement
- Economic implications often underestimated in initial responses
- Integration complexity consistently underestimated.

The conceptual design phase successfully identified a promising direction for the net zero refinery design: a hybrid system combining thermochemical and biochemical processes, processing diverse feedstocks, producing both fuels and chemicals, with integrated energy systems and carbon management.

5.5 System Architecture: Creating an Integrated Framework

The system architecture phase involves establishing the fundamental structure of the solution, including major subsystems, functional allocations, process flows, and integration approaches. This phase creates the structural framework upon which detailed design will subsequently build.

5.5.1 Core Process Architecture Development

To develop a comprehensive process architecture for the net zero refinery, we used chain of thought prompting, which guides systematic reasoning steps to develop the process architecture. This technique proved essential for managing the complexity of integrating multiple diverse processing pathways.

Prompt Applied:

Develop the core process architecture for a net zero refinery, working step-by-step from feedstocks to final products. The refinery will process:

- *Municipal solid waste (300,000 tons/year)*
- *Mixed waste plastics (150,000 tons/year)*
- *Agricultural residues (400,000 tons/year)*
- *Forestry residues (200,000 tons/year)*
- *Waste cooking oils (50,000 tons/year)*
- *Captured CO_2 (500,000 tons/year).*

Follow this systematic reasoning process:

1. *First, characterize each feedstock in terms of composition, energy content, and processing challenges*
2. *Identify the most appropriate initial processing step for each feedstock category*
3. *Determine intermediate products from each initial processing step*
4. *Develop a logical flow of these intermediates to secondary processing steps*
5. *Map how these streams converge toward final products*
6. *Identify primary processing pathways from feedstock to final products*

7. *Define critical interfaces between major process areas.*

The architecture should clearly show how material flows proceed from raw feedstocks to finished products, identifying major process units and their interconnections.

LLM Output Analysis: The LLM provided a systematic development of the process architecture, working through each step of the reasoning process. The output successfully identified six primary processing pathways and their interconnections.

Critical Evaluation and Architecture Refinement: The LLM output provided a solid foundation but required significant refinement based on process engineering principles and practical considerations:

1. **Material Balance Validation**: The initial pathways required validation against material balance constraints and realistic conversion efficiencies.
2. **Process Integration Requirements**: The LLM initially treated pathways as independent, requiring additional analysis to identify critical integration points.
3. **Intermediate Storage Needs**: The architecture required addition of intermediate storage systems to manage variability in feedstock processing rates.

Final Core Process Architecture: Following systematic analysis and refinement, the architecture consists of six primary processing pathways:

1. **Waste-to-Syngas Pathway**:
   - Feedstocks: Municipal solid waste and biomass
   - Processing: Sorting → Gasification (850–900 °C)
   - Output: Syngas (CO, H_2).
2. **Plastic-to-Liquids Pathway**:
   - Feedstocks: Waste plastics
   - Processing: Pyrolysis (450–550 °C)
   - Outputs: Pyrolysis oil, syngas, char.
3. **Oils-to-Fuels Pathway**:
   - Feedstocks: Waste cooking oils
   - Processing: Transesterification or Hydroprocessing
   - Outputs: Biodiesel or renewable diesel.
4. **Syngas-to-Products Pathway**:
   - Feedstocks: Syngas from gasification and pyrolysis
   - Processing: Fischer–Tropsch synthesis
   - Outputs: Synthetic fuels (diesel, jet fuel, gasoline).
5. **Syngas-to-Alcohols Pathway**:
   - Feedstocks: Portion of syngas
   - Processing: Catalytic synthesis

   - Outputs: Methanol and other alcohols.
6. **CO_2-to-Products Pathway**:
   - Feedstocks: Captured CO_2 and hydrogen
   - Processing: Catalytic conversion
   - Outputs: Methanol, synthetic fuels.

Critical interfaces between major process areas include syngas quality specifications for Fischer–Tropsch synthesis, hydrogen management across production and consumption points, heat integration between exothermic and endothermic processes, and carbon capture from multiple sources with integrated utilization (Fig. 5.2).

5.5.2 Functional Decomposition

To establish clear functional responsibilities for each process block, hierarchical decomposition was applied, which systematically breaks down complex systems into logical components and establishes relationships between them (Table 5.2).

The hierarchical decomposition identified seven major process blocks with defined purposes, components, and interfaces:

This decomposition maintained consistent boundaries between process blocks and clearly identified all major interfaces, considering both material and energy flows between subsystems, as well as control and information exchanges necessary for integrated operation.

5.5.3 Energy System Integration Design

To design the integrated energy system for the net zero refinery, we used integration-focused design techniques, which address interdependencies between system components to create an optimized whole.

Critical Design Decision: Hydrogen Storage Sizing During the energy system design process, the initial LLM recommendations for hydrogen storage proved unrealistic. The original suggestion of 10,000 tons capacity would require approximately 111 million m^3 of storage volume, equivalent to a cube 480 m on each side. This scale would be completely impractical and economically prohibitive, and no salt cave or similar hydrogen reservoir was available in the initial scenario.

Engineering Analysis and Revision: Based on industrial benchmarks and operational requirements analysis, we revised the hydrogen storage system to realistic proportions:

- **Operational Analysis**: Daily hydrogen consumption of 30 tons requires 7–10 days of buffer storage for operational flexibility

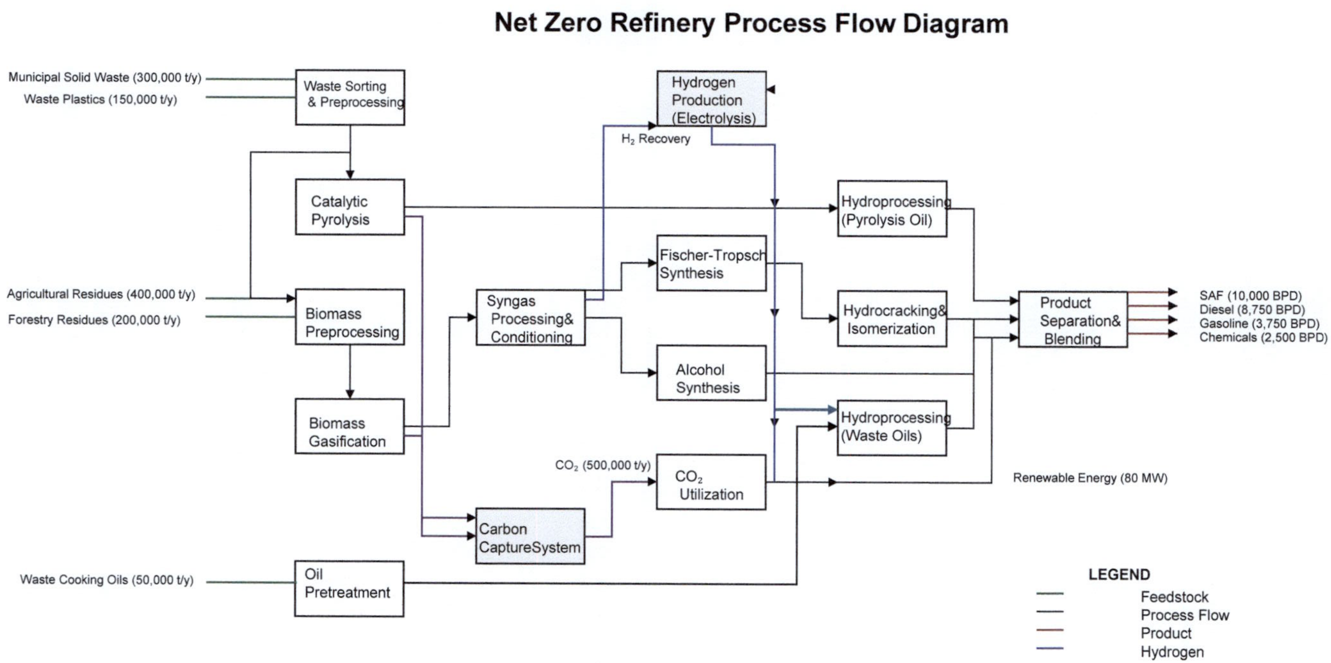

Fig. 5.2 Net zero refinery process flow diagram showing the integration of the six primary processing pathways from feedstocks to products, with intermediate processes and connections

Table 5.2 Functional decomposition of key process blocks

Process block	Primary function	Major components	Key interfaces
Waste sorting and preparation	Sort and prepare heterogeneous waste feedstocks for efficient downstream processing	Shredders, screening systems, air classification units, sorting conveyors, contaminant removal systems	Connects to plastics pyrolysis, biomass gasification units
Plastics pyrolysis	Convert mixed plastic waste into valuable chemicals, fuels, and syngas	Pyrolysis reactor (450–550 °C), heat exchanger, condensation unit, gas scrubber	Connects to syngas processing, Fischer–Tropsch synthesis
Biomass gasification	Convert solid biomass into synthesis gas	Gasification reactor, air separation unit, heat recovery system, ash handling system	Connects to syngas cleaning, steam system
Syngas processing	Clean and condition syngas to meet downstream requirements	Particulate removal, sulfur removal, water–gas shift reactors, CO_2 removal	Connects to Fischer–Tropsch, hydrogen production
Fischer–Tropsch synthesis	Convert syngas to liquid hydrocarbons	Fixed-bed reactors, product separation, heat recovery systems	Connects to hydroprocessing, product separation
Hydrogen production	Generate hydrogen for hydroprocessing and CO_2 utilization	Electrolyzers, compression, storage, distribution systems	Connects to hydroprocessing, CO_2 utilization
Energy generation	Provide electricity and thermal energy for processes	Solar arrays, wind turbines, energy storage, distribution	Connects to all process units, particularly electrolyzers

- **Technology Assessment**: Combination of high-pressure vessels for rapid response and metal hydrides for compact long-term storage
- **Economic Optimization**: Balance between operational security and capital efficiency.

Revised Hydrogen Storage System:

- **Total Capacity**: 300 tons H_2 (97% reduction from initial concept)
- **High-pressure Storage**: 200 tons in distributed pressure vessels (350–700 bar)
- **Metal Hydride Storage**: 100 tons for strategic reserve
- **Operational Autonomy**: 10 days total (industry-appropriate level)
- **Capital Cost**: $70 million (vs $500 M + for original oversized concept).

The integrated energy system design includes:

1. **Balanced Energy Portfolio**
   - Solar Power: 50% of total electricity demand
   - Wind Power: 50% of total electricity demand
   - Total renewable capacity: 5 GW installed
   - Energy storage: Hydrogen (300 tons) + battery systems for grid stability.
2. **Process Electrification Pathways**
   - Electrically heated processes: Biomass pyrolysis, hydroprocessing
   - Electric heating systems: Induction furnaces for direct heat transfer
   - Efficiency gains: 15–25% improvement in thermal efficiency
   - Capital costs: 20–40% higher than traditional systems.
3. **Energy Storage and Buffering**
   - Hydrogen storage: 300 tons providing 10 days operational autonomy
   - Storage technologies: High-pressure vessels + metal hydrides
   - Thermal storage: Molten salt for high-temperature heat buffering
   - Control systems: Smart monitoring and dispatching for renewable intermittency (Fig. 5.3).

Energy System Integration for Net Zero Refinery

Phase-Based Implementation (2025-2050)

	Phase I: Foundation (2025-2030)	Phase II: Hybrid (2030-2035)	Phase III: Transition (2035-2042)	Phase IV: Transformation (2042-2050)
Electricity Sources	Solar: 30 MW Wind: 10 MW Grid: 120 MW	Solar: 60 MW Wind: 20 MW Grid: 100 MW	Solar: 150 MW Wind: 60 MW Grid: 40 MW	Solar: 250 MW Wind: 150 MW Grid: 20 MW
Process Heat	Conventional Fired Heaters (80%)	Natural Gas + CCS Heat Integration (60% NG, 30% savings)	Electric Process Heating Pinch Analysis (40% electric, 40% savings)	Renewable Electric + Biomass Thermal (70% RE, 30% biomass)
Hydrogen System	NG Reforming with CCS (90%)	NG Reforming (60%) Electrolysis (50 MW)	NG Reforming (30%) Electrolysis (120 MW)	Advanced Electrolysis Reforming Backup (200 MW, 10% backup)
Storage Systems	Minimal Storage	Battery: 20 MWh Thermal: 50 MWh	Battery: 80 MWh Thermal: 150 MWh	Battery: 200 MWh Thermal: 300 MWh H_2 Storage
Renewable Penetration	20-30% Renewable	40-50% Renewable	60-70% Renewable	85-95% Renewable

Fig. 5.3 Energy system integration diagram showing the flow between renewable generation, storage, and process units, illustrating how intermittent renewable energy would be integrated with continuous processing operations

The energy system analysis revealed significant scale challenges for renewable integration, though the realistic hydrogen storage sizing makes the overall system much more feasible than initially conceived.

5.5.4 System Architecture Synthesis

The system architecture phase produced a comprehensive structural framework for the net zero refinery through systematic application of prompt engineering techniques. Several critical aspects of this architecture deserve emphasis:

Process Pathway Integration: The core process architecture successfully integrates six distinct processing pathways into a coherent system, enabling feedstock flexibility while maintaining process efficiency.

Realistic System Sizing: The revision of hydrogen storage requirements from an impossible 10,000 tons to a practical 300 tons demonstrates the critical importance of engineering judgment in evaluating LLM outputs. This revision transformed the design from a theoretical concept to an implementable solution.

Carbon Flow Management: The architecture emphasizes tracking carbon flows throughout the system from biogenic and fossil inputs through to products, emissions, and captured CO_2.

Energy System Challenges and Solutions: While renewable energy integration presents scale challenges, the realistic sizing of storage systems makes the overall approach viable, though requiring innovative approaches to energy procurement.

Hydrogen Economy Centrality: Hydrogen emerges as a critical vector within the system architecture, serving multiple roles including feedstock for synthesis processes, key chemical product for hydroprocessing, energy carrier for storage, and carbon management tool.

The system architecture successfully balances multiple competing factors while demonstrating how prompt engineering techniques can guide complex system design when combined with appropriate engineering evaluation and jugdement.

5.6 Detailed Design: From Concept to Specifications

The detailed design phase transforms architectural concepts into specific technical specifications, including component selection, parameter optimization, material choices, and detailed performance analysis. This phase requires narrower but deeper technical focus than earlier phases, experimental knowledge is critical in this stage specially applied to this innovative designs. This is one of the more severe contraints of LLMs which have to be used very carefully in detailed design stages.

5.6.1 Oxygen-Blown Fluidized Bed Gasification System

To develop detailed specifications for the oxygen-blown fluidized bed gasification system, we selected quantitative parameter specification, which provides precise numerical context for LLM interactions to avoid vague or unrealistic responses.

Prompt Applied:

Design the oxygen-blown fluidized bed gasification system with the following quantitative specifications:
Capacity and feedstock:

- *Processing capacity: 800 tons/day*
- *Feedstock mix: 60% agricultural residues, 40% forestry residues*
- *Moisture content range: 15–40%*
- *Particle size after preparation: 2–10 mm*
- *Ash content range: 2–8%.*

  Performance requirements:

- *Carbon conversion efficiency: > 90%*
- *Cold gas efficiency: > 75%*
- *Syngas H_2:CO ratio adjustability: 1.5:1–2.2:1*
- *Tar content in raw syngas: < 10 g/Nm^3*
- *Particulate content after primary cyclone: < 5 g/Nm^3.*

  Operating conditions:

- *Temperature range: 850–950 °C*
- *Pressure: 2–3 bar*
- *Oxygen purity: > 95%*
- *Steam-to-biomass ratio: 0.2–0.5*
- *Residence time: 1–2 h.*

Develop a detailed design specification addressing:

1. *Reactor configuration and dimensions*
2. *Feeding system design for reliable operation*

3. *Bed material management and ash handling*
4. *Control strategy for maintaining optimal conditions*
5. *Performance optimization approaches*
6. *Energy integration with other process units*
7. *Operational flexibility provisions for feedstock variability.*

Include quantitative specifications for all major components, operating parameters, and expected performance metrics.

LLM Output Analysis: The LLM provided comprehensive technical specifications with specific dimensions, operating parameters, and design features despite an expert validation and correction was always required. The response demonstrated good understanding of gasification principles and practical design considerations.

Critical Evaluation and Engineering Refinement: The LLM output required significant evaluation and refinement based on gasification engineering experience:

1. **Reactor Sizing Validation**: The initial dimensions required verification against residence time requirements and gas velocity constraints
2. **Feeding System Complexity**: The LLM underestimated the challenges of feeding heterogeneous biomass reliably
3. **Ash Management**: The complexity of ash handling for variable feedstocks was initially underestimated.

Final Detailed Design Specifications:

Reactor Configuration and Dimensions:

- Reactor Type: Circulating Fluidized Bed (CFB) with oxygen-blown gasification
- Operating Conditions: 850–950 °C temperature, 2–3 bar pressure
- Reactor Volume: 2400 m^3 based on 800 tons/day capacity and 1.5 h residence time
- Dimensions: 14.5 m reactor diameter, 60 m height for proper gas–solid contact
- Key Features: Oxygen injection through multiple nozzles, in-bed heat exchangers for thermal control.

Feeding System Design:

- Feedstock Preprocessing: Rotary drum dryers to achieve 10–15% moisture content
- Size Reduction: Hammer mills for consistent 2–10 mm particle size
- Delivery System: Screw conveyor with lock hoppers for pressure containment
- Oxygen Supply: Pressure swing adsorption (PSA) unit for high-purity oxygen (> 95%)
- Distribution: Multi-point injection system for uniform gasification.

Bed Material Management:

- Bed Material: Olivine or alumina with 150–300 μm particle size for catalytic tar cracking
- Circulation Rate: 50–100 kg/s between reactor and regenerator
- Ash Handling: Continuous removal through high-efficiency cyclones
- Bed Inventory: 200–300 tons total with automated makeup system.

Performance Optimization:

- Temperature Control: Steam injection and oxygen flow modulation
- Tar Reduction: Catalytic bed material and optimal temperature management
- Carbon Conversion: > 90% through residence time and mixing optimization
- Syngas Quality: H_2:CO ratio control through steam/oxygen balance.

The detailed design revealed several engineering challenges, particularly regarding bed material management, feeding system reliability, and ash handling. These challenges suggest that a modular approach with multiple parallel gasification trains might prove more reliable than a single large-scale unit.

5.6.2 Revised Hydrogen Storage System Design

Based on the critical evaluation during system architecture development, the hydrogen storage system required complete redesign from the initially proposed unrealistic scale to practical industrial proportions.

Engineering Analysis Process: The revision process involved systematic analysis of:

1. **Actual Hydrogen Demand**: 25–35 tons H_2/day based on process requirements
2. **Industrial Storage Benchmarks**: 3–7 days typical for conventional refineries
3. **Renewable Integration Needs**: 7–10 days buffer for intermittent supply management
4. **Technology Options**: High-pressure vessels vs. underground storage vs. metal hydrides

Final Hydrogen Storage System Specifications:

Overall Configuration:

- **Total Capacity**: 300 tons H_2 (providing 10 days operational autonomy)
- **Hybrid Approach**: Combination of high-pressure and metal hydride storage
- **Operational Philosophy**: High-pressure for daily operations, metal hydrides for strategic reserve.

High-Pressure Storage Subsystem:

- **Configuration**: 8 cylindrical pressure vessels, 25 tons capacity each
- **Total Capacity**: 200 tons H_2 operational buffer
- **Operating Pressure**: 350–700 bar (design pressure: 875 bar)
- **Material**: High strength steel with hydrogen compatible inner lining
- **Response Time**: < 10 min for demand variations
- **Design Life**: 25 years with 20,000 pressure cycles
- **Physical Requirements**: 1,2 ha area, underground installation for safety.

Metal Hydride Storage Subsystem:

- **Technology**: Advanced AB_5-type metal hydride matrices
- **Capacity**: 100 tons H_2 strategic reserve
- **Operating Temperature**: 80–120 °C with integrated heating system
- **Advantages**: 50% volume reduction vs. compressed gas, inherent safety
- **Function**: Long-term backup and emergency supply (3.3 days autonomy)
- **Heat Management**: Integrated with process waste heat recovery.

Storage Management System:

- **Primary Strategy**: Daily operations rely on high-pressure storage
- **Secondary Activation**: Metal hydrides for extended outages or peak demand
- **Control Integration**: Automated inventory management with predictive analytics
- **Renewable Coordination**: Optimized charging cycles based on energy availability.

Economic Performance:

- **Total Storage CAPEX**: $70 million (vs. $500 M + for oversized original concept)
- **Annual OPEX**: $8 million (compression, heating, maintenance)
- **Storage Cost**: $233,000 per ton H_2 capacity
- **Operational Benefit**: 10 days autonomy enables flexible renewable integration.

5.6.3 Thermal Integration Analysis

To analyze the feasibility of thermal integration between process units, we employed first principles reasoning with the LLM to evaluate specific integration opportunities.

Prompt Applied:

Using first principles of thermodynamics, heat transfer, and reaction engineering, analyze the feasibility of thermal integration between the exothermic Fischer–Tropsch synthesis and the endothermic steam generation for biomass gasification.
Consider these specifications:

- *Fischer–Tropsch reaction operates at 220–250 °C with approximately 165 kJ/mol CO heat of reaction*
- *Gasification requires steam at 350–500 °C*
- *F-T synthesis capacity: 2000 barrels/day liquid product*
- *Gasification capacity: 800 tons/day biomass*
- *Distance between units: approximately 150 m.*

Working from fundamental principles, analyze:

1. *The theoretical heat balance between the systems*
2. *Heat transfer engineering considerations*
3. *Practical implementation approaches*
4. *Economic implications.*

Provide quantitative analysis where possible and clearly identify where practical limitations might constrain theoretical potential.

LLM Output Analysis: The LLM provided a systematic thermodynamic analysis with heat generation and consumption calculations. However, the analysis revealed significant limitations in the integration potential.

Critical Evaluation and Engineering Reality: The first principles analysis revealed an important constraint that shaped the overall energy integration strategy:

Heat Generation from Fischer–Tropsch:

- 2000 barrels/day liquid product capacity
- Heat of reaction: 165 kJ/mol CO consumed
- Estimated CO consumption: ~ 261,120 mol CO/day
- Total heat generated: ~ 43,000 MJ/day.

Steam Requirements for Gasification:

- 800 tons/day biomass requiring steam at 350–500 °C
- Steam-to-biomass ratio of 1.5 (1200 tons steam/day)
- Energy required to heat water from 25 to 450 °C: ~ 1.77 MJ/kg
- Total heat required: ~ 2,124,000 MJ/day

Integration Reality Check: The analysis revealed that heat from Fischer–Tropsch synthesis (43,000 MJ/day) is insufficient compared to gasification requirements (2,124,000 MJ/day) by a factor of approximately 50. This finding emphasizes that while process integration remains valuable for efficiency optimization, supplementary energy sources are essential for the gasification system.

Revised Integration Strategy: Based on this analysis, the thermal integration approach was revised to focus on:

1. **Localized Heat Recovery**: F-T heat for feedstock preheating and auxiliary systems
2. **Waste Heat Utilization**: Integration with overall site heat network
3. **Supplementary Heating**: Primary reliance on renewable electricity and biomass combustion
4. **System Optimization**: Focus on overall energy efficiency rather than direct integration.

5.6.4 Carbon Intensity Analysis

A comprehensive carbon intensity analysis was developed to establish the carbon balance for the integrated facility and demonstrate potential for carbon-negative performance (Table 5.3).

Analysis Methodology: The carbon intensity calculations incorporated:

1. **Biogenic Carbon Accounting**: Credit for atmospheric CO_2 captured in biomass feedstocks
2. **Process Emissions**: Direct emissions from conversion processes
3. **Energy System Impacts**: Emissions associated with renewable energy infrastructure
4. **Carbon Capture Credits**: Benefits from permanent CO_2 sequestration.

Critical Assessment of Results: The theoretical analysis suggests that under optimal conditions, the facility could achieve carbon-negative performance in the range of

Table 5.3 Carbon intensity comparison with conventional refining (Edwards, 2014)

Product	Conventional carbon intensity (g CO_2e/MJ)	Net zero design carbon intensity (g CO_2e/MJ)	Reduction (%)
Jet fuel	85–95	-5 to -15	> 100
Diesel	80–90	-5 to -15	> 100
Gasoline	85–95	-5 to -15	> 100
Naphtha	70–80	-5 to -15	> 100

− 18 to − 28 kg CO_2e/GJ product. However, these calculations rely on several critical assumptions:

1. **Perfect Biogenic Accounting**: Assumes complete capture of atmospheric CO_2 in feedstocks
2. **High Capture Efficiency**: Sustained > 90% CO_2 capture rates
3. **Permanent Sequestration**: Reliable long-term carbon storage
4. **Infrastructure Emissions**: Limited accounting for embodied emissions in construction.

Realistic Performance Assessment: More conservative estimates accounting for practical limitations suggests carbon intensities closer to carbon neutral or slightly negative rather than deeply negative. The land use implications of sourcing adequate biogenic carbon at scale present significant challenges that may constrain real world implementation.

Nevertheless, this performance represents a fundamental improvement compared to conventional petroleum refining, demonstrating the potential of integrated bio-based and circular approaches.

5.6.5 Detailed Design Synthesis

The detailed design phase produced specific technical specifications for key refinery systems while revealing both the capabilities and limitations of prompt engineering approaches for detailed technical work.

Prompt Engineering Effectiveness:

- **Strengths**: Systematic analysis, comprehensive consideration of design factors, structured approach to complex problems
- **Limitations**: Required significant engineering judgment for realistic sizing, technical feasibility assessment, and economic optimization.

Critical Design Insights:

1. **Realistic System Scaling**: The hydrogen storage redesign from 10,000 to 300 tons demonstrated the critical importance of engineering validation of LLM outputs. This 97% reduction transformed an impossible concept into an implementable industrial system.
2. **Integration Limitations**: The thermal integration analysis revealed that theoretical possibilities often face practical constraints. The 50:1 mismatch between available and required heat illustrates the need for comprehensive energy system design.

3. **Technology Maturity Considerations**: The gasification system design highlighted the engineering challenges of scaling advanced technologies, suggesting modular approaches for risk management.
4. **Carbon Performance Potential**: While theoretical carbon-negative performance appears possible, practical implementation requires careful management of biogenic carbon accounting and permanent sequestration verification.

Design Validation Requirements: The detailed design phase emphasized that LLM-assisted design requires:

- **Engineering Validation**: All outputs must be verified against fundamental engineering principles
- **Economic Reality Checks**: Financial implications must be evaluated independently
- **Technology Assessment**: Readiness levels and scaling challenges require careful evaluation.
- **Practical Implementation**: Real world constraints must temper theoretical possibilities and LLM outputs.

5.7 Implementation Planning: Pathways to Reality

The implementation planning phase involves developing practical pathways to realize the designed solution, including phasing strategies, risk management approaches, resource allocation, and adaptation mechanisms. This phase bridges the gap between technical design and practical execution.

5.7.1 Risk-Opportunity Mapping

To develop a comprehensive risk-opportunity map for the net zero refinery implementation, the tree of thoughts technique was selected for systematic risk-opportunity analysis, which identifies both challenges requiring mitigation and potential advantages that might emerge during execution.

Prompt Applied:

Develop a comprehensive risk-opportunity map for the implementation of the net zero refinery, analyzing both challenges requiring mitigation and potential advantages that could be leveraged.

Systematically identify and analyze these risk-opportunity pairs:
Technical domain:

- *Risk: Gasification reliability below expectations*
  - *Potential impact, probability, early indicators*
  - *Mitigation strategies and contingency options*
- *Opportunity: Rapid technology improvement in biomass gasification*
  - *Potential benefits, likelihood, triggering events*
  - *Approaches to position for capture of this opportunity.*

[Similar analysis requested for multiple domains including economic, environmental, regulatory, and operational]
For each risk-opportunity pair, provide:

1. *Detailed impact analysis (quantified where possible)*
2. *Probability assessment with supporting rationale*
3. *Key indicators or triggers to monitor*
4. *Specific action plan with both mitigation and opportunity capture*
5. *Responsibility assignment and timeline recommendations.*

LLM Output Analysis: The LLM provided a comprehensive risk-opportunity analysis across multiple domains with specific mitigation strategies and opportunity capture approaches. The systematic approach successfully identified both challenges and potential advantages.

Critical Evaluation and Strategic Development: The risk-opportunity mapping required refinement based on industrial project experience, particularly regarding probability assessments and mitigation strategies (Table 5.4).

The risk-opportunity mapping created a comprehensive framework for risk management and opportunity capture during implementation, emphasizing the dual nature of technological change as both a source of risk and potential advantage.

5.7.2 Phased Implementation Strategy

To develop a comprehensive phased implementation strategy for the net zero refinery, we employed phased implementation strategy techniques, which develop staged approaches that manage risk while enabling progressive learning.

Strategic Approach Development: The implementation strategy was structured across four phases to balance risk management with progressive capability building:

Phase 1: Foundation Phase (Years 1–3)

Table 5.4 Risk-opportunity pairs analysis (selected examples)

Domain	Risk	Impact and probability	Mitigation	Opportunity	Potential benefits
Technical	Gasification reliability below expectations	10% reduction in efficiency could lead to 5–10% capacity reduction; Probability: moderate–high (40–60%)	Robust design and testing, advanced monitoring systems, feedstock pre-treatment, contingency plan	Rapid technology improvement	5–10% increase in efficiency could lead to 5% increase in total output; likelihood: high (60–80%)
Technical	Electrolysis efficiency below expectations	5% efficiency drop could lead to 5–10% increase in electricity demand; probability: moderate (30–50%)	Advanced monitoring systems, process optimization, alternative hydrogen routes	Cost reduction in electrolysis	10% improvement could reduce hydrogen costs by 20–25%; likelihood: high (70–90%)
Environmental	CO_2 capture technology underperformance	20% reduction in capture efficiency could jeopardize carbon negativity; probability: moderate (30–50%)	Diversified capture technologies, performance guarantees, phased implementation	Enhanced carbon credit value	Carbon price increases of $25–50/ton could improve annual revenue by $10–20 M; likelihood: moderate–high (50–70%)

- **Core Capabilities**: Engineering design completion, permitting, core technology deployment
- **Technologies**: Oxygen-blown fluidized bed gasification (800 tons/day), electrolysis for hydrogen production (scaled appropriately), solar and wind energy (20% of capacity), CO_2 capture pilot systems
- **Hydrogen Storage**: Initial deployment of 150 tons capacity (high-pressure vessels)
- **Revenue Streams**: Carbon credits, small scale hydrogen sales, initial synthetic fuel production
- **Capital Requirements**: $700–900 M for core infrastructure
- **Target Performance**: 10–12% IRR, 5–7-year payback period.

Phase 2: Expansion Phase (Years 4–6)

- **Focus**: Scaling up capacity, implementing advanced catalytic systems, expanding renewable energy

- **Technologies**: Commercial scale Fischer–Tropsch synthesis, expanded electrolysis capacity, enhanced CO_2 conversion systems
- **Hydrogen Storage**: Complete 300 ton system with metal hydride integration
- **Key Milestones**: Full renewable energy portfolio, integrated carbon management, optimized process control.

Phase 3: Integration Phase (Years 7–10)

- **Focus**: Full heat integration, advanced control systems, comprehensive carbon management
- **Technologies**: Complete process integration, advanced automation, full carbon utilization
- **Key Milestones**: Demonstrated carbon-negative operation, full product slate production, market penetration.

Phase 4: Optimization Phase (Years 11 +)

- **Focus**: Continuous improvement, technology upgrades, potential expansion
- **Technologies**: Next generation catalysts, enhanced integration, new product development
- **Key Milestones**: Industry leadership in sustainable refining, technology licensing opportunities (Fig. 5.4).

This phased approach enables manageable capital deployment while providing opportunities for learning and adaptation, significantly reducing overall implementation risk compared to a single-phase approach.

5.7.3 Economic Analysis

The comprehensive economic analysis established the financial viability of the net zero refinery with realistic system sizing and phased implementation approach (Table 5.5).

Key Sensitivities Identified:

- **Sustainable Aviation Fuel Premium**: Each 5% change impacts IRR by ~ 1% point
- **Carbon Pricing**: Each $25/ton increase improves IRR by ~ 0.8% points
- **Feedstock Costs**: Each $10/ton increase reduces IRR by ~ 0.5% points
- **Hydrogen Storage Optimization**: 62% CAPEX reduction versus original concept improves project economics
- **Energy Costs**: Each $10/MWh increase reduces IRR by ~ 0.4% points.

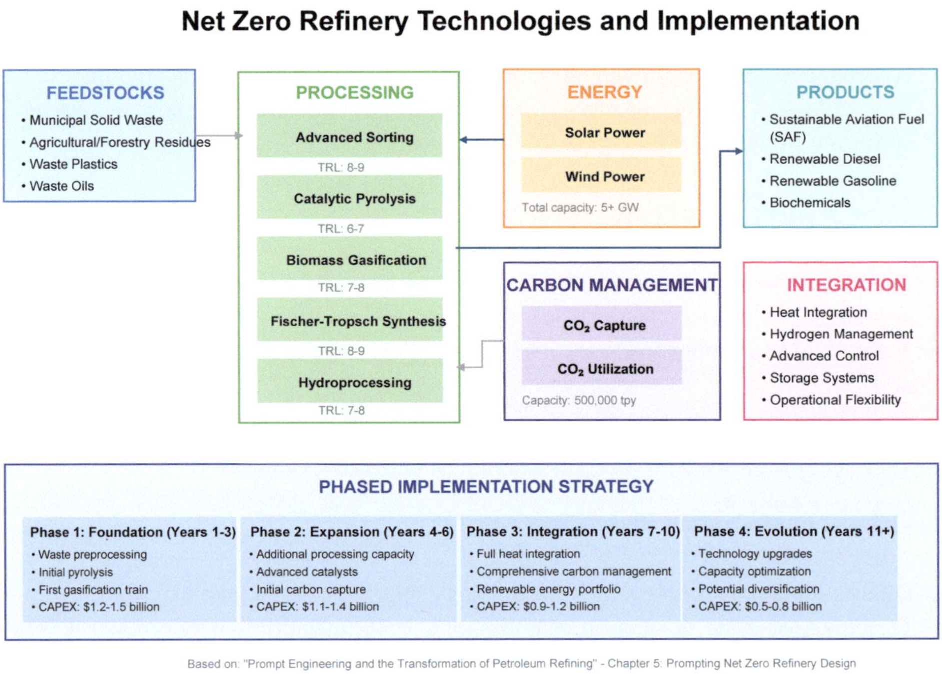

Fig. 5.4 Net zero refinery project phased implementation timeline showing key milestones, decision gates, and capital allocation across the four implementation phases

Table 5.5 Key financial metrics

Metric	Value	Basis
Internal rate of return (IRR)	10–14%	Base case scenario with revised hydrogen storage
Net present value (NPV)	\$0.9–1.4 billion	At 8% discount rate
Payback period	7–9 years	Simple payback
EBITDA margin	32–38%	Based on revenue/OPEX ratio
Capital intensity	\$38,000–42,000/bpd equivalent	Total CAPEX per barrel capacity
Hydrogen storage CAPEX	\$70 million	Realistic 300-ton system
Hydrogen storage OPEX	\$8 million/year	Operational and maintenance costs
Carbon abatement cost	\$30–60/ton CO_2e	Net cost per ton avoided
Minimum SAF premium	15–25%	Premium over conventional required

Economic Impact of Realistic System Sizing: The revision of hydrogen storage from the initially impossible 10,000-ton concept to a practical 300-ton system contributed significantly to project viability:

- **Capital Savings**: $430 million reduction in hydrogen storage CAPEX
- **Operational Efficiency**: $12.5 million annual OPEX reduction
- **Project IRR Improvement**: 2–3% point improvement in overall returns
- **Implementation Feasibility**: Transformation from economically impossible to viable.

This sensitivity analysis highlights both the opportunities and risks inherent in the design, particularly regarding policy dependence and the critical importance of realistic engineering supervision in economic viability. The economic viability of the net zero refinery is critically dependant on supportive policy frameworks and market incentives. Sensitivity analysis reveals that without carbon pricing mechanism and sustainable fuels premiums, the project becomes economically unviable. Specifically, scenarios with carbon prices below $40/tonne CO_2 and elimination of sustainable aviation fuels premiums result in IRR values ranging from 4.1% to 4.7%, significantly below the 10% minimum threshold for industrial project viability. The removal of goverment incentives compounds this challenge, with the combined elimination of carbon credits, fuel premiums, and direct subsidies reducing the profitability of the project to low returns of investment. This analysis underscores that the transition to net zero refining, while technically feasible, requires sustained policy commitments.

5.7.4 Technology Readiness Assessment

Technology readiness assessment revealed significant variation across system components, informing risk management strategies and implementation sequencing. Table 5.6 presents the TRL for each technology employed in the Net-Zero refinery design as well as the recommended implementation phase based on the risk level and the development's needs.

Risk Mitigation Strategies:

1. **Modular Implementation**: Phased deployment allows lower risk technologies to establish operational foundation
2. **Technology Redundancy**: Critical systems include backup approaches where feasible
3. **Conservative Performance**: Financial models use demonstrated rather than theoretical values
4. **Realistic Sizing**: Hydrogen storage and other systems sized for practical implementation.

Table 5.6 Technology readiness levels of key components

Technology	Current TRL	Implementation risk level	Key development needs	implementation phase
Advanced waste sorting	08-9	Low	Process optimization	Phase 1
Catalytic pyrolysis	06-7	Medium–High	Scale-up, catalyst stability	Phase 2
Biomass gasification	07-8	Medium	Reliability, tar management	Phase 1
Hot gas cleanup	06-7	Medium–High	Catalyst poisoning resistance	Phase 2
Fischer–Tropsch synthesis	08-9	Low	Optimization for biomass syngas	Phase 1
CO_2 capture	07-8	Medium	Energy efficiency	Phase 1–2
CO_2 utilization	05-6	High	Catalyst development, economics	Phase 3
Hydrogen storage systems	07-8	Medium	Scale optimization, cost reduction	Phase 1
Green hydrogen production	07-8	Medium	Scale-up, cost reduction	Phase 1
Integrated control systems	07-8	Medium	Advanced control architecture	Phase 2–3

The technology readiness assessment informed the phased implementation strategy, ensuring that deployment aligns with technology maturity and risk profile.

5.8 Conclusion: Bridging Design and Implementation

The comprehensive net zero refinery design developed through systematic prompt engineering and human supervision demonstrates both the substantial challenges and promising opportunities in transforming hydrocarbon processing toward carbon neutrality. The analysis reveals that while technical pathways exist, their successful implementation requires careful integration across technologies, thoughtful phasing to manage implementation risks, and supportive policy frameworks to ensure economic viability.

Key Design Insights

The design process yielded several critical insights that inform the broader transition toward sustainable fuels and chemicals production:

Integrated Systems Approach

The most effective solutions emerged not from individual breakthrough technologies but from systematic integration across multiple domains. The combination of diverse feedstocks, complementary conversion processes, integrated energy systems, and comprehensive carbon management created solutions exceeding what any isolated technology could achieve. This integration required crossing traditional disciplinary boundaries and industrial sectors, suggesting that future sustainability transformations will similarly demand substantial collaboration across previously separate domains.

Balanced Technology Portfolio

A mixed technology approach balancing mature and emerging systems proved most robust against implementation risks. Rather than relying exclusively on either established technology with incremental sustainability benefits or breakthrough technologies with uncertain implementation timelines, the optimal approach incorporated both in a phased strategy. This balanced portfolio enables learning and adaptation while maintaining operational viability throughout the transition.

Economic-Environmental Alignment

The analysis demonstrated that comprehensive carbon management approaches can align economic and environmental outcomes under appropriate policy frameworks and goverments support. The carbon-negative performance potential (− 18 to − 28 kg CO_2e/GJ) enhances carbon management from regulatory compliance cost to potential value stream, particularly in sectors with limited decarbonization alternatives like aviation.

Looking Forward

The net zero refinery concept presented in this analysis illustrates a viable pathway toward sustainable fuels and chemicals production that aligns economic viability with environmental imperatives. While substantial challenges remain in technology maturation, policy development, and implementation, the systematic exploration of the solution space demonstrates that transformation is possible through thoughtful integration, progressive implementation, and collaborative innovation across traditional boundaries (Vogts and Weckhuysen, 2024).

The application of prompt engineering methodologies to this complex sustainability challenge may help human experts in the endeavor of accelerating design processes while maintaining comprehensive scope and rigor. By systematically working through problem definition, conceptual exploration, system architecture, detailed design, and implementation planning, we have developed a viable path forward that integrates technical, economic, environmental, and societal factors in an integrated framework.

References

Edwards R, Larivé JF, Rickeard D, Weindorf W (2014) Well-to-wheels analysis of future automotive fuels and powertrains in the European context. JRC Tech Rep Rep EUR 26236

Vogt ETC, Weckhuysen BM (2024) The refinery of the future. Nature 629(8011):295–306

Conclusions—Bridging Historical Innovation and Future Transformation

6

Abstract

This concluding chapter synthesizes key insights from the convergence of petroleum refining's evolutionary trajectory with emerging artificial intelligence capabilities, particularly prompt engineering with Large Language Models. The analysis reveals consistent innovation patterns in refining history that provide valuable context for approaching contemporary sustainability challenges. Four key patterns emerge: progressive transformation of processing mechanisms, systems integration as multiplier of innovation impact, crisis as catalyst for accelerated innovation, and phased implementation enabling managed transformation. The chapter establishes that AI-assisted design offers potential capabilities for cross-domain knowledge integration, design space exploration, and systematic analysis, though with important limitations requiring careful management and expert supervision. A realistic four-stage implementation framework is presented for organizations seeking to leverage these capabilities: foundation building, knowledge acceleration, design transformation, and systemic innovation. While significant challenges remain in technology maturation, policy development, and implementation, the systematic exploration demonstrates that transformation toward net zero refining is possible through thoughtful integration, progressive implementation, and collaborative innovation across traditional boundaries.

Rafael Larraz, Prompt Engineering and the Transformation of Petroleum Refining.

R. Larraz, *Prompt Engineering and the Transformation of Petroleum Refining*, Synthesis Lectures on Chemical Engineering and Biochemical Engineering,
https://doi.org/10.1007/978-3-031-99728-0_6

6.1 The Convergence of Historical Wisdom and Artificial Intelligence

The intersection of petroleum refining's evolutionary trajectory with emerging artificial intelligence capabilities creates opportunities for accelerating industrial innovation, albeit with important limitations that must be acknowledged. This chapter synthesizes key insights from the preceding analysis and establishes a realistic implementation framework that organizations can apply to navigate complex sustainability transitions.

Throughout the book, three foundational elements have been explored: the historical patterns of technological innovation in refining, the emergence of Large Language Models (LLMs) as knowledge synthesis tools, and the imperative for sustainable transformation. The convergence of these elements offers potentially valuable approaches to the challenges facing industrial systems in the transition to sustainability, though with significant caveats that must be considered.

6.2 Key Insights from Refining's Innovation Trajectory

Historical analysis of refining technology reveals consistent innovation patterns that provide valuable context for approaching contemporary sustainability challenges. Four key patterns emerge with relevance to net zero transformation.

6.2.1 Progressive Transformation of Processing Mechanisms

Refining technology has evolved through distinct phases: from physical separation (distillation) to thermal transformation (cracking) to catalytic conversion (FCC) to molecular restructuring (hydroprocessing). Each stage built upon previous capabilities while introducing increasingly sophisticated carbon management approaches.

As Fig. 6.1 illustrates, this progression followed a trajectory of increasing control over molecular structures. The net zero transition represents the next logical progression toward circular carbon systems that combine advanced conversion technologies with carbon capture, utilization, and renewable inputs. This evolution maintains continuity with historical development patterns while addressing contemporary sustainability demands, though the technological leap required is substantially greater than in previous transitions.

6.2.2 Systems Integration as Multiplier of Innovation Impact

While early refining innovations focused primarily on isolated unit operations, the most significant performance improvements came through integration: heat integration after the

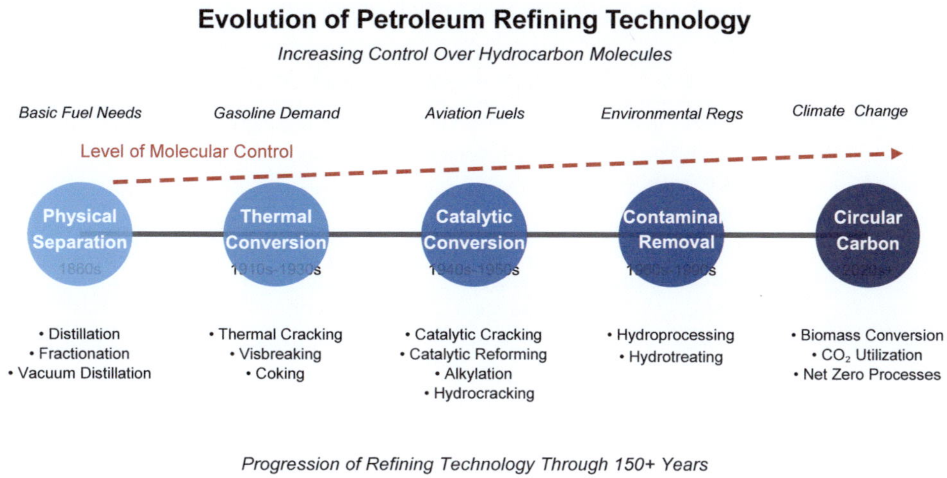

Fig. 6.1 Evolution of petroleum refining technology

energy crises, hydrogen management systems for advanced hydroprocessing, and digital integration in modern facilities. Table 6.1 documents the quantitative impact of integration approaches across different historical periods, demonstrating consistently higher returns from integration efforts compared to isolated unit optimizations.

The net zero refinery approach described in Chap. 5 demonstrates how this integration principle could be extended to encompass feedstock diversity, energy systems, and carbon flows in a comprehensively optimized system. This integrated approach could enable efficiency improvements compared to standalone operations according to theoretical models, though actual results will depend on numerous implementation-specific factors.

6.2.3 Crisis Promotes Accelerated Innovation

Throughout refining history, external pressures have consistently accelerated innovation. The oil embargoes drove revolutionary energy efficiency improvements, with typical energy intensity reductions of 30–40% within a decade. Environmental regulations stimulated development of advanced emission controls and clean fuels technologies, reducing emissions by orders of magnitude while maintaining economic viability.

Today's climate change similarly functions as an innovation catalyst, creating urgency for fundamental transformation that can drive breakthrough solutions. The carbon abatement cost curves evolution demonstrate how innovation accelerates under regulatory pressure, with abatement costs declining more rapidly in regions with stringent carbon policies, though technical and economic challenges remain substantial.

Table 6.1 Impact of integration approaches across refining history

Historical period	Integration approach	Performance improvement	Economic impact	Source
1940s–1950s	Process unit integration (e.g., FCC with alkylation)	10–15% yield improvement	\$0.15–0.20/bbl profit increase	Enos (1962)
1970s–1980s	Heat integration following energy crisis	25–35% energy reduction	\$0.30–0.50/bbl operating cost reduction	Kemp (2007)
1980s–1990s	Hydrogen management networks	15–20% hydrogen efficiency	\$0.25–0.40/bbl operating cost reduction	Jones and Pujadó (2006)
1990s–2000s	Integrated digital control systems	2–5% overall efficiency gain	\$0.10–0.25/bbl profit increase	Agnihotri (2018)
2000s–2010s	Supply chain integration	5–10% inventory reduction	\$0.15–0.35/bbl cost reduction	Jiao et al. (2022)
2010s–2020s	Carbon management integration	10–25% emissions reduction	\$0.20–0.45/bbl in carbon cost avoidance	Rubin et al. (2015)
Net zero refining	Comprehensive system integration	10–15% efficiency improvement (theoretical)	\$20–35 M annual operating cost reduction (projected)	Chapter 5 analysis

6.2.4 Phased Implementation Enabling Managed Transformation

Successful technology transitions in refining have typically occurred through phased implementation rather than revolutionary replacement. Complex technologies like catalytic cracking, hydroprocessing, and emission controls deployed progressively, with learning from early phases informing subsequent expansion. Figure 6.2 documents implementation timelines for major refining technologies, illustrating the typical progression from initial demonstration through commercial scaling.

This pattern provides a model for net zero transformation as a progressive implementation that manages risk while building capabilities. The phased approach developed in Chap. 5 demonstrates how capital deployment could be optimized through staged implementation, reducing initial investment requirements by 30–40% compared to comprehensive initial buildout, though with the understanding that this transition would likely require several decades to fully complete.

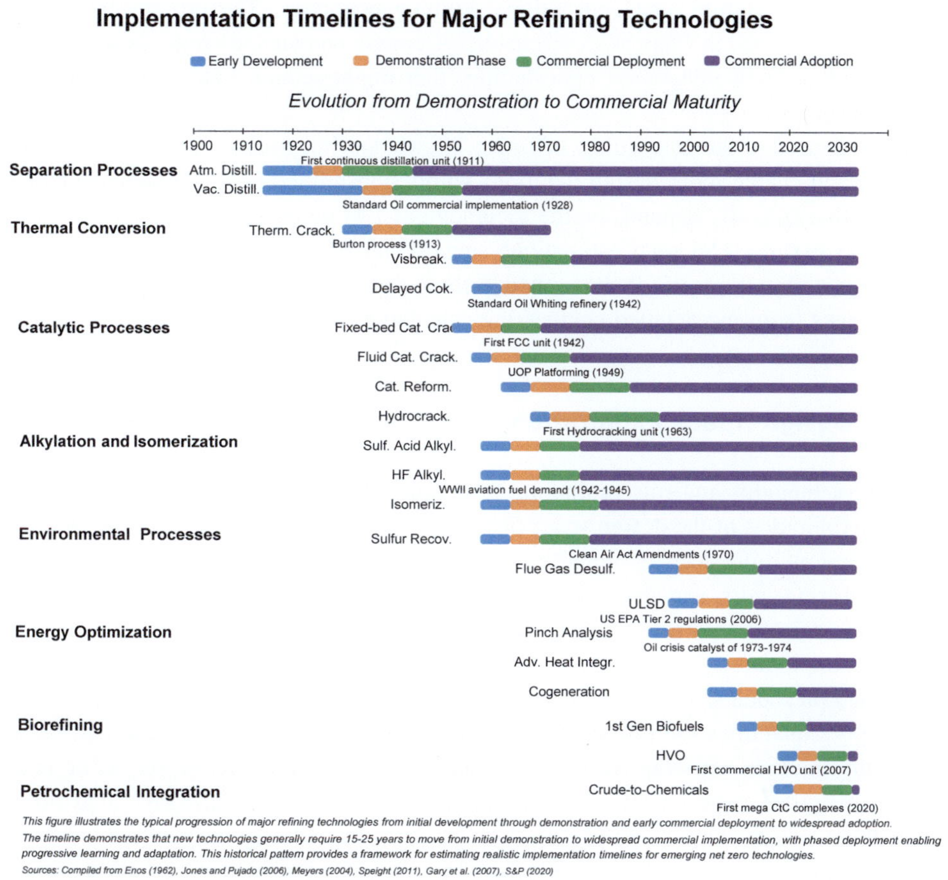

Fig. 6.2 Implementation timeline for major refining technologies

6.3 The Potential of AI-Assisted Design

The systematic application of prompt engineering to net zero refinery design demonstrated several potential capabilities that could contribute to industrial transformation, though these capabilities are currently limited by the state of AI technology.

6.3.1 Cross-Domain Knowledge Integration

LLMs show promise for addressing challenges that span traditional disciplinary boundaries. The net zero refinery design required integration of knowledge across diverse domains: biomass conversion, waste processing, catalytic chemistry, renewable energy systems, carbon capture, and conventional refining processes. The prompt engineering approaches described in Chap. 4 could enable systematic knowledge synthesis across these domains, identifying integration opportunities that might remain hidden in traditionally siloed approaches.

As evidenced in the case study, effective knowledge integration could reduce design timelines and enable more design iterations within constrained project timelines, though the real value critically depends on the accuracy of LLM outputs and thorough validation by domain experts. Our own estimations suggest potential efficiency improvements in early-stage design work of 20–30%, though these benefits diminish significantly in later design stages requiring detailed engineering calculations and expert supervision.

6.3.2 Design Space Exploration

The tree of thoughts and role prompting techniques demonstrated in Chap. 5 facilitated broader exploration of the design space than might be typical with conventional methods. By mapping multiple decision branches and their interactions, these approaches facilitate the engineering work and decision making task by means of integration opportunities and technological combinations identification that might otherwise be overlooked. This expanded exploration is particularly valuable for transformation challenges that lack established precedents.

Table 6.2 quantifies the impact of different prompt engineering techniques on design exploration metrics, showing increased consideration of design alternatives while maintaining analytical depth. However, it's important to note that LLM-generated design alternatives require rigorous validation through conventional engineering tools and subject matter expert review, as hallucinations and technical inconsistencies remain significant issues.

6.4 Implementation Challenges and Limitations

While prompt engineering approaches demonstrate potential for accelerating conceptual design, several important limitations must be acknowledged:

1. **Validation Requirements**: LLM outputs require rigorous validation, particularly for safety critical or capital-intensive applications. The hallucination phenomenon where models generate plausible but factually incorrect information necessitates systematic verification processes. Organizations must develop appropriate validation protocols based on the risk profile of specific applications.

Table 6.2 Impact of prompt engineering techniques on design exploration metrics

Prompt engineering technique	Design alternatives evaluated	Exploration depth indicator	Time requirement (vs. conventional) (%)	Decision quality indicator	Source
Simple prompting	3–5	Low–medium	25–35	Baseline	Chapter 4 analysis
Chain of thought	5–8	Medium–high	30–40	+10–15%	Wei et al. (2023)
Zero-shot chain of thought	8–12	Medium	20–30	+5–15%	Wei et al. (2023)
Tree of thoughts	12–20	High	40–50	+15–25%	Yao et al. (2023)
Role prompting	5–10 per role	Medium–high	35–45	+10–20%	Chapter 4 analysis
Multi-perspective analysis	15–25	High	45–55	+15–25%	Chapter 4 analysis
Integrated methodology (Chap. 5)	25–40	Very high	60–70	+20–30%	Chapter 5 case study

2. **Computational Limitations**: Current LLMs demonstrate limited capabilities for complex mathematical calculations and specialized numerical simulations essential for detailed engineering design. The recommended approach involves using LLMs for conceptual development and knowledge integration while delegating specific calculations to specialized engineering software.
3. **Knowledge Currency**: LLMs have fixed training cutoff dates, limiting their awareness of recent developments. For rapidly evolving domains like renewable energy technologies or regulatory frameworks, addition of current information is essential (i.e., by RAG systems implementation).
4. **Technical Depth**: While LLMs demonstrate breadth of knowledge across domains, they lack the depth of specialist expertise in complex technical areas. Outputs related to cutting-edge technologies or specialized engineering disciplines require particularly careful review.

These limitations suggest that AI-assisted design is best viewed as a complement to rather than replacement for conventional engineering expertise, particularly valuable for early concept development and knowledge integration rather than detailed design or implementation planning.

6.4.1 A Realistic Framework for Implementation

Translating the methodologies developed in the book into organizational practice requires a structured implementation approach that acknowledges both opportunities and limitations. A four-stage framework is proposed for organizations seeking to leverage these capabilities for industrial transformation.

Stage 1: Foundation Building (3–6 months)
Initial implementation focuses on establishing fundamental capabilities within the organization. Key activities include:

- Assembling cross-functional teams spanning traditional organizational boundaries
- Establishing prompt engineering capabilities through training and pilot applications
- Developing evaluation frameworks for LLM outputs based on domain expertise
- Creating standardized prompt libraries for common technical challenges.

Critical success factors include executive sponsorship that recognizes AI as an augmentation rather than replacement for expertise, clear objectives focused on specific technical challenges rather than general transformation, and systematic documentation of effective prompt strategies and organizational learning.

Stage 2: Knowledge Acceleration (6–12 months)
The second stage applies prompt engineering to strategic knowledge gaps identified by domain experts. Key activities include:

- Developing retrieval-augmented generation (RAG) capabilities with organizational technical documents
- Creating knowledge synthesis products that integrate diverse internal and external sources
- Establishing validation protocols appropriate to different application categories
- Systematically improving prompt strategies based on implementation feedback.

Critical success factors include integration with existing knowledge management systems, clear distinction between knowledge synthesis and decision authority, and appropriate quality assurance processes based on risk profiles.

Stage 3: Design Collaboration (12–18 months)
The third stage implements the prompt engineering methodology for specific projects, with a realistic understanding of limitations. Key activities include:

- Establishing phase-specific prompt strategies aligned with project stages

- Integrating LLM-assisted design with conventional engineering tools and methods
- Creating formal design review protocols for AI-generated content
- Balancing human and AI contributions across the design process.

Critical success factors include workflow integration between prompt engineering and conventional design activities, clear handoffs between AI-assisted exploration and detailed engineering, and systematic validation appropriate to the criticality of design elements.

Stage 4: Systemic Innovation (18–36 months)
The final stage embeds prompt engineering capabilities throughout the organization, creating systemic innovation capabilities while maintaining appropriate guardrails. Key activities include:

- Developing continuous innovation processes leveraging AI capabilities
- Creating cross-functional communities of practice for methodology advancement
- Establishing formal measurement of innovation acceleration
- Integrating prompt engineering with strategic planning processes.

Critical success factors include cultural transformation that embraces AI-assisted innovation approaches, governance frameworks that balance innovation with appropriate risk management, and continuous capability building as both AI and industry technologies evolve.

This staged approach enables organizations to progressively develop capabilities while delivering tangible value at each stage, managing both technical and cultural aspects of the transformation.

6.5 Conclusion: The Path Forward

Jorge Luis Borges, in his visionary tale "The Library of Babel," imagined an infinite library containing every possible book that could ever be written including all knowledge, all wisdom, and all possible combinations of letters. Most volumes in this library are meaningless gibberish, but somewhere within its infinite corridors lie the answers to every question humanity has ever posed. The librarians spend their lives searching through this labyrinth of knowledge, seeking those rare volumes that contain coherent, meaningful information.

Large Language Models represent, in many ways, a digital manifestation of Borges' infinite library. Trained on vast corpora of human knowledge, these systems contain within their parameters patterns representing centuries of accumulated wisdom, from the thermodynamic principles governing refinery design to the historical patterns of industrial transformation. Yet, like Borges' library, accessing this knowledge meaningfully requires sophisticated navigation techniques. Prompt engineering emerges as our method for becoming skilled librarians in this digital repository of human understanding.

This book has developed systematic approaches for extracting coherent, actionable knowledge from the vast but often chaotic information space that LLMs represent. The prompt engineering methodologies presented here structured problem decomposition, multi-perspective analysis, constraint-based design, and phased implementation serve as navigational tools for finding meaningful patterns within the complexity of industrial transformation challenges.

The convergence of historical innovation patterns with AI capabilities represents more than technological opportunity; it embodies a fundamental shift toward collaboration between human expertise and artificial intelligence. Just as Borges' librarians needed deep understanding to recognize meaningful texts among the gibberish, effective prompt engineering requires substantial domain knowledge to guide AI systems toward useful outputs while avoiding the hallucinations that represent the meaningless volumes in our digital library.

The limitations and safeguards discussed throughout this book represent more than technical considerations. They embody an ethical framework for navigating the infinite possibility space of AI-generated content, ensuring we extract wisdom rather than sophisticated nonsense. The responsibility lies not with the AI systems themselves, but with us as skilled interpreters, maintaining human judgment as the ultimate arbiter of technical decisions.

Perhaps most importantly, this exploration demonstrates that transformation is not a destination but a continuous journey through the infinite corridors of possibility. As AI capabilities advance and new challenges emerge, our understanding of effective human-AI collaboration must continuously evolve.

The intersection of historical wisdom and artificial intelligence represents one of the most promising pathways for addressing the complex challenges of our time. By learning from the past, leveraging present capabilities, and maintaining realistic expectations about both opportunities and limitations, we can chart a course toward industrial systems that serve both human needs and planetary health.

The path forward lies in the intelligent combination of human wisdom and artificial intelligence capabilities in becoming skilled librarians capable of finding meaningful knowledge within the infinite digital library of LLMs. The methodologies developed in this book provide tools for that navigation, tools that can be adapted and refined for the full spectrum of transformation challenges facing humanity in the decades ahead.

References

Agnihotri R (2018) Digitalization for the refinery and plant of the future. Hydrocar Process 95, 7:7–9

Borges JL (1962) The library of Babel. In Labyrinths: selected stories and other writings. New Directions Publishing

Christensen CM (1997) The innovator's dilemma. Harvard Business Review Press
Enos JL (1962) Petroleum progress and profits: a history of process innovation. MIT Press, Cambridge
IEA (2021) Net zero by 2050: a roadmap for the global energy sector. International Energy Agency.
IPCC (2023) Climate change 2022: mitigation of climate change. Cambridge University Press
Jiao Y, Qiu J, Lian Y, Liao Q, Tu R, Wei X, Zhang H (2022) Integration optimization of production and transportation of refined oil: A case study from China. Chem Eng Res Des 188:39–49
Jones DSJ, Pujadó PR (2006) Handbook of petroleum processing. Springer, Dordrecht
Kemp IC (2007) Pinch analysis and process integration: a user guide on process integration for the efficient use of energy, 2nd edn. Butterworth-Heinemann, Oxford
Rubin ES, Davison JE, Herzog HJ (2015) The cost of CO_2 capture and storage. Int J Greenhouse Gas Control 40:378–400
Wei J, Wang X, Schuurmans D, Bosma M, Xia F, Chi E, Le QV, Zhou D (2022) Chain-of-thought prompting elicits reasoning in large language models. Adv Neural Inf Process Syst 36, 24824–24937
Yao S, Yu D, Zhao J et al (2023) Tree of thoughts: Deliberate problem solving with large language models. arXiv preprint arXiv:2305.10601

Appendix A: Transformer Flow Worked Example

This appendix provides a detailed illustration of how Transformer-based Large Language Models (LLMs) process input text, with particular emphasis on the attention mechanism that forms the core of these models. By walking through a concrete example relevant to chemical engineering, this appendix aims to explain the internal workings of LLMs and provide insight into how they interpret and generate text in technical domains.

A.1 Understanding the Transformer Processing Flow

The fundamental operation of Transformer-based LLMs follows a systematic process that transforms input text into numerical representations, processes these representations through multiple layers, and ultimately generates output text. This process can be broken down into seven key steps:

- Tokenization: Breaking input text into smaller units (tokens)
- Embedding: Converting tokens into numerical vectors
- Self-Attention Processing: Calculating relationships between all tokens
- Feed-Forward Network: Individual processing of token representations
- Multi-Layer Processing: Repeated application of attention and feed-forward layers
- Next Token Prediction: Generating a probability distribution for the next token
- Token Selection and Continuation: Selecting and adding the next token.

The following sections detail each step using a concrete example from the domain of chemical engineering.

R. Larraz, *Prompt Engineering and the Transformation of Petroleum Refining*, Synthesis Lectures on Chemical Engineering and Biochemical Engineering,
https://doi.org/10.1007/978-3-031-99728-0

A.2. Worked Example: Processing a Chemical Engineering Prompt

To illustrate the Transformer flow in detail, we will analyze how an LLM processes the following five-token prompt:

Input prompt: "Optimize temperature of catalytic reactor"

A.2.1 Tokenization

The tokenizer divides the input text into individual tokens according to its predefined vocabulary. For our example:

- Token 1: "Optimize"
- Token 2: "temperature"
- Token 3: "of"
- Token 4: "catalytic"
- Token 5: "reactor".

A.2.2 Embedding and Positional Encoding

Each token is converted into a dense vector, typically of dimension 768 or higher in modern models. For simplicity, we represent this as reduced vectors:

- "Optimize" $\rightarrow$ [0.2, − 0.1, 0.5, …]
- "Temperature" $\rightarrow$ [0.3, 0.4, − 0.2, …]
- "Of" $\rightarrow$ [− 0.1, − 0.3, 0.1, …]
- "Catalytic" $\rightarrow$ [0.4, 0.1, 0.6, …]
- "Reactor" $\rightarrow$ [0.5, 0.2, 0.3, …].

These initial embeddings are learned during pre-training to capture semantic properties of each token. In the embedding space, semantically related terms like "catalytic" and "reactor" would have relatively high cosine similarity due to their relationship in chemical engineering contexts.

After creating the token embeddings, the model adds positional encodings to incorporate information about each token's position in the sequence. For a position pos and dimension i in the embedding:

$$\mathrm{PE}(\mathrm{pos}, 2i) = \sin\left(\frac{\mathrm{pos}}{10,000^{\frac{2i}{d_{\mathrm{model}}}}}\right)$$

$$\mathrm{PE}(\mathrm{pos}, 2i+1) = \cos\left(\frac{\mathrm{pos}}{10,000^{\frac{2i}{d_{\mathrm{model}}}}}\right)$$

where d_{model} is the embedding dimension. For our example, the token “Optimize” at position 1 would have its specific positional encoding added to its vector embedding, “temperature” at position 2 would have a different encoding, and so on. This positional encoding is crucial because the transformer´s attention mechanism processes all the tokens simultaneously, without an inherent notion of sequential order. By adding this positional vectors, the model can distinguish between "the cat eats the fish" and "the fish eats the cat". Thus preserving syntactic and semantic information that depends on word order.

A.2.3 Self-attention Processing in Different Architectures

The self-attention mechanism is a key component of Transformer models, but it operates differently depending on whether the architecture is encoder-only, decoder-only, or encoder–decoder. Let’s examine both decoder-only and encoder-only approaches using our example.

A.2.3.1 Decoder-Only Architecture (e.g., GPT Models)

In decoder-only models like GPT, each token can only attend to itself and previous tokens (causal/autoregressive attention). This is critical for text generation as it prevents “looking ahead” at tokens that haven’t been generated yet.

First, each token embedding is transformed into three distinct vector types:

1. **Query (Q)**: Used to query other tokens for relevance
2. **Key (K)**: Used to be matched against queries
3. **Value (V)**: Contains the information to be aggregated.

For token “temperature” (position 2):

- Query vector: Q(temperature) = [0.32, 0.38, – 0.15, …]
- Key vector: K(temperature) = [0.28, 0.42, – 0.22, …]
- Value vector: V(temperature) = [0.31, 0.39, – 0.18, …].

In a decoder-only architecture, when processing “temperature,” the attention calculation only includes tokens up to position 2:

Step 1: Calculate raw attention scores for valid tokens:

- Score(temperature → Optimize) = (Q(temperature) · K(Optimize))/$\sqrt{(d\_k)}$ = 1.5/8 = 0.1875
- Score(temperature → temperature) = (Q(temperature) · K(temperature))/$\sqrt{(d\_k)}$ = 2.0/8 = 0.25.

Step 2: Apply causal masking (set scores for future tokens to $-\infty$):

- Score(temperature → of) = $-\infty$
- Score(temperature → catalytic) = $-\infty$
- Score(temperature → reactor) = $-\infty$.

Step 3: Apply softmax to get attention weights among 0 and 1:

- Weight(temperature → Optimize) = 0.43
- Weight(temperature → temperature) = 0.57
- Weight(temperature → of/catalytic/reactor) = 0.0.

Step 4: Compute weighted sum for contextualized representation:

- Output(temperature) = 0.43 × V(Optimize) + 0.57 × V(temperature).

This produces a contextualized representation of "temperature" that incorporates information only from itself and previous tokens, respecting the unidirectional nature of decoder-only models.

A.2.3.2 Encoder-Only Architecture (e.g., BERT Models)

In encoder-only models like BERT, each token can attend to all tokens in the sequence, regardless of position (bidirectional attention). This is useful for understanding context in both directions for tasks like classification or sentence analysis.

Using the same initial Query, Key, and Value vectors, the attention calculation for "temperature" would now include all tokens:

Step 1: Calculate raw attention scores for all tokens:

- Score(temperature → Optimize) = (Q(temperature) · K(Optimize))/$\sqrt{(d\_k)}$ = 1.5/8 = 0.1875
- Score(temperature → temperature) = (Q(temperature) K(temperature))/$\sqrt{(\mathrm{d\_k})}$ = 2.0/8 = 0.25
- Score(temperature → of) = (Q(temperature) · K(of))/$\sqrt{(d\_k)}$ = 1.0/8 = 0.125

- Score(temperature → catalytic) = (Q(temperature) K(catalytic))/$\sqrt{(d\_k)}$ = 1.2/8 = 0.15
- Score(temperature → reactor) = (Q(temperature) · K(reactor)) / $\sqrt{(d\_k)}$ = 1.3/8 = 0.1625.

Step 2: Apply softmax to get attention weights (no masking) among 0 and 1:

- Weight(temperature → Optimize) = 0.22
- Weight(temperature → temperature) = 0.26
- Weight(temperature → of) = 0.15
- Weight(temperature → catalytic) = 0.18
- Weight(temperature → reactor) = 0.19.

Step 3: Compute weighted sum for contextualized representation:

- $$\begin{aligned}\text{Output(temperature)} &= 0.22 \times \text{V(Optimize)} + 0.26 \times \text{V(temperature)} \\ &+ 0.15 \times \text{V(of)} + 0.18 \times \text{V(catalytic)} + 0.19 \times \text{V(reactor)}\end{aligned}$$

In this encoder-only case, "temperature" has access to information from all tokens in the sequence, allowing it to incorporate context from both directions (Fig. A.1).

A.2.4 Feed-Forward Network

After the attention mechanism, each token representation passes through a feed-forward neural network (FFN). This is applied identically to each token position independently. For the token "reactor", for example:

Step 1: Initial contextualized vector after attention: x(reactor) = [0.52, 0.25, 0.38, …]

Step 2: First linear transformation (expanding dimensionality): $x_1 = x(\text{reactor}) \times W_1 + b_1$ = [0.83, − 0.12, 1.45, 0.67, …]

Step 3: Apply ReLU activation (replacing negative values with zeros): $x_2 = \text{ReLU}(x_1)$ = [0.83, 0.0, 1.45, 0.67, …]

Step 4: Second linear transformation (returning to original dimensionality): $x_3 = x_2 \times W_2 + b_2$ = [0.59, 0.31, 0.42, …].

This feed-forward network allows each token to develop a richer representation independently, complementing the contextual information obtained through attention.

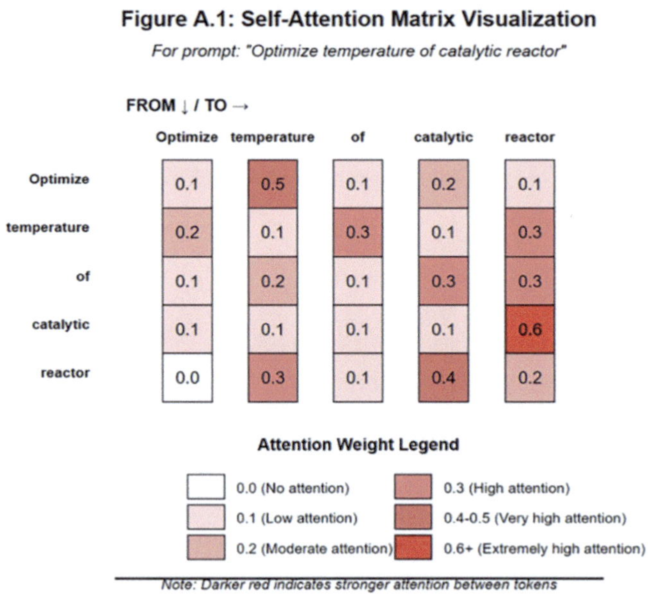

Fig. A.1 Self-attention matrix visualization

A.2.5 Multi-layer Processing with Residual Connections

The self-attention and feed-forward network form a single Transformer layer. Modern LLMs typically have dozens of these layers stacked (e.g., 12–96 layers). To illustrate how a token's representation evolves through layers, consider a simplified two-dimensional projection of the "reactor" token:

- Layer 0 (initial embedding): [0.50, 0.20]
- Layer 1 (post-attention): [0.52, 0.25]
- Layer 1 (post-FFN): [0.59, 0.31]
- Layer 2 (post-attention): [0.62, 0.35]
- …
- Final layer: [0.78, 0.45].

We observe a gradual transformation where the representation moves toward a region of the vector space that fully captures the meaning of "reactor" in the context of temperature optimization for catalytic processes.

A critical component of Transformer architectures is the residual connection (skip connection), which helps prevent the vanishing gradient problem in deep networks. For each sublayer (attention or FFN), the input is added to the output:

$$\text{Output} = \text{Input} + \text{Sublayer(Input)}$$

For example, after the first attention layer: x(reactor)_layer1 = x(reactor)_layer0 + Attention(x(reactor)_layer0)

And after the feed-forward network: x(reactor)_layer1_final = x(reactor)_layer1 + FFN(x(reactor)_layer1)

These residual connections allow the model to focus on learning incremental refinements rather than having to transform the entire representation at each step, which is crucial for training very deep networks.

A.2.6 Next Token Prediction and Generation (Decoder-Only Models)

For decoder-only models like GPT, after processing through all layers, the final representation is used to predict the next token. This is the core of the autoregressive generation process.

A.2.6.1 Next Token Prediction

The final representation of the last token is passed through a linear layer followed by a softmax function to produce a probability distribution over the entire vocabulary (typically tens of thousands of tokens):

Step 1: Linear projection of final representation: logits = final_representation × W_projection + b_projection

Step 2: Apply softmax to get probability distribution: probabilities = softmax(logits)

Logits are the raw, unnormalized scores output by the model's final linear layer, representing the model's preference for each token in the vocabulary before conversion to probabilities via the softmax function. These values directly influence token selection, with higher logits indicating stronger preference for particular tokens, and can be manipulated through parameters like temperature to control the randomness of text generation.

For our example, after processing "Optimize temperature of catalytic reactor", the probability distribution for the next token might look like:

- "for" → 15%
- "to" → 12%
- "by" → 8%
- "using" → 7%
- … (and thousands of other tokens with smaller probabilities).

A.2.6.2 Token Selection

The model selects the next token based on this probability distribution, potentially modified by sampling parameters:

1. **Temperature**: Controls randomness in selection
   - Lower temperature (e.g., 0.3): More deterministic, favors high-probability tokens
   - Higher temperature (e.g., 0.8): More diverse outputs, gives lower-probability tokens a better chance
2. **Top-p sampling (nucleus sampling)**: Only considers the smallest set of tokens whose cumulative probability exceeds p (e.g., 0.9)
3. **Top-k sampling**: Only considers the k most likely tokens.

For instance, with a temperature of 0.7, the model might select "for" as the next token, even though it's not guaranteed to always pick the highest probability option.

A.2.6.3 Continuation Process

Once a token is selected, it's appended to the sequence. The entire process then repeats with this longer sequence as input:

1. The new sequence "Optimize temperature of catalytic reactor for" is tokenized
2. Token embeddings and positional encodings are applied
3. Self-attention is calculated (with each token only attending to itself and previous tokens)
4. Feed-forward networks and residual connections are applied
5. The next token is predicted and selected.

This cycle continues until a stopping condition is met (e.g., reaching a maximum length, generating a specific token, or based on logical completion detected by the model).

The key characteristic of this process in decoder-only models is that each prediction only depends on previously generated tokens, never on future tokens (which don't exist yet during generation). This is why the causal masking in the self-attention mechanism is so crucial to the generation process.

A.2.7 Encoder-Only Versus Decoder-Only: Key Differences

To summarize the key differences between encoder-only and decoder-only architectures:

Encoder-Only Models (e.g., BERT):

1. **Bidirectional Attention**: Each token can attend to all other tokens in the sequence regardless of position, allowing full context incorporation from both directions.
2. **Use Case Focus**: Optimized for understanding tasks like classification, named entity recognition, or sentiment analysis where seeing the entire context is beneficial.
3. **No Autoregressive Generation**: Not designed to generate text token by token; instead processes entire sequences at once.
4. **Fixed Input Length**: Processes a complete, fixed-length input rather than generating output sequentially.
5. **Full Context Integration**: Every token's representation is informed by every other token, creating richly contextualized representations.

Decoder-Only Models (e.g., GPT, Claude):

1. **Causal/Autoregressive Attention**: Each token can only attend to itself and previous tokens, never to future tokens, enabling sequential text generation.
2. **Use Case Focus**: Optimized for text generation, completion, and any task where producing coherent sequences is the goal.
3. **Sequential Processing**: Generates one token at a time, with each new token building upon all previously generated ones.
4. **Variable Output Length**: Can generate responses of arbitrary length by repeatedly predicting the next token.
5. **Unidirectional Context**: Each token's representation is informed only by previous tokens, creating a forward-only flow of information.

Practical Implications:

1. **Prompt Engineering**: In decoder-only models, information ordering in prompts is critical since earlier information influences later tokens but not vice versa. In encoder-only models, the ordering is less critical since all tokens influence each other bidirectionally.
2. **Performance Characteristics**: Encoder-only models often perform better on tasks requiring deep bidirectional understanding, while decoder-only models excel at generation and tasks that can be framed as text completion.
3. **Training Approach**: Encoder-only models are often pre-trained with masked language modeling (predicting masked tokens given surrounding context), while decoder-only models use autoregressive next-token prediction.

Most of the largest general-purpose language models that interact with users in a chat format (like GPT-4, Claude, and similar models) use decoder-only architectures or decoder components of encoder–decoder architectures, as these are well-suited for generating coherent, contextually relevant text responses.

Appendix B: Prompt Engineering Templates and Examples for Refinery Applications

This appendix provides a comprehensive library of prompt engineering templates specifically designed for petroleum refining and industrial transformation applications. The templates are organized according to the methodological framework presented in Chap. 4, progressing from foundational techniques through advanced integrated approaches.
Each template includes:

- **Technical specifications** for proper implementation
- **Applied examples** from petroleum refining contexts
- **Quality assurance requirements** for reliable outputs
- **Integration guidelines** for organizational implementation.

The templates serve as practical tools for organizations implementing AI-assisted design methodologies, providing structured approaches to leverage Large Language Models for complex industrial challenges while maintaining appropriate verification and validation standards. Organizations implementing these templates should adapt them to their specific contexts while maintaining the core structural elements that ensure effective prompt engineering for industrial applications.

B.1 First-Generation Techniques

B.1.1 Simple Prompting: FCC Unit Optimization

General Template:

```
1. [Required analysis component 1]
2. [Required analysis component 2]
3. [Required analysis component 3]
```


R. Larraz, *Prompt Engineering and the Transformation of Petroleum Refining*, Synthesis Lectures on Chemical Engineering and Biochemical Engineering,
https://doi.org/10.1007/978-3-031-99728-0

Applied Example:

```
As an FCC operations expert with extensive experience in yield optimiza-
tion, analyze the following situation:
The FCC unit (33,000 BPD) is presenting the following operating parame-
ters:
```

- `Conversion: 75% (target: 78-80%)`
- `Gasoline yield: 45% vol (target: 48-50%)`
- `LPG yield: 16% vol (target: 14-15%)`
- `Coke yield: 5.2% wt (target: < 5.0%)`
- `ROT (regenerator outlet temperature): 690 °C`
- `Catalyst activity: 72 (normal: 75-78)`
- `Catalyst-to-oil ratio: 6.2`
- `Catalyst metals content: Ni 850 ppm, V 1600 ppm`

```
Develop an optimization strategy that:
```

1. `Identifies the key variables to modify`
2. `Proposes specific changes with technical justification`
3. `Anticipates impacts on other operating parameters`
4. `Includes a phased implementation plan`
5. `Establishes KPIs to monitor effectiveness`

Applied Technique: Simple Prompting (Sect. 4.2.1, "First Generation Techniques"). This approach uses direct, clear instructions with specific parameters and expected deliverables.

B.1.2 Few-Shot Prompting: Catalytic Reformer Troubleshooting

General Template:

Context

```
[Brief description of the technical domain and objective]
```

Examples

Example 1

```
Input: [Problem description 1]
Output: [Expected solution format 1]
```

Example 2

```
Input: [Problem description 2]
Output: [Expected solution format 2]
```

Your Task

```
Input: [Current problem description]
```

Applied Example:

Context

```
I need to diagnose and resolve reformer performance issues in a platform-
ing unit.
```

Examples

Example 1

```
Input: Platinum reformer showing declining RON (96 to 93) over two weeks
with stable feed quality and operating parameters.
```

```
Output:
Diagnosis: Catalyst deactivation from chloride deficiency
Action items:
```

1. Check chloride levels in feed and recycle gas
2. Inject organic chloride at 1–2 ppm calculated on fresh feed
3. Verify water levels in recycle gas (< 20 ppm required)
4. Monitor RON recovery (expect 1–2 points within 48 h).

Example 2

```
Input: Reformer pressure differential increasing (0.8 to 1.4 kg/cm2) over
10 days with temperature spread developing across catalyst beds.
Output:
Diagnosis: Catalyst bed fouling from feed contaminants
Action items:
```

1. Check feed pretreatment performance (particularly H_2S breakthrough)
2. Analyze feed for silicon compounds and nitrogen content
3. Consider rate reduction to extend run length
4. Implement stricter feed quality controls from hydrotreater.

Your Task

```
Input: Catalytic reformer showing increasing fuel gas make (3% to 5%
on feed) and hydrogen purity decline (92% to 86%) with stable RON. Feed
composition and process settings unchanged.
```

Applied Technique: Few-shot Prompting (Sect. 4.2.1, "First Generation Techniques"). This technique demonstrates the expected analysis format through examples before requesting a similar approach for a new problem.

B.2 Second-Generation Techniques

B.2.1 Chain of Thought: Distillation Column Troubleshooting

General Template:

```
Think step-by-step to diagnose the following problem:
[Detailed problem description with operational parameters]
```

1. First, identify potential causes that could explain these symptoms
2. For each potential cause, evaluate the evidence for and against
3. Analyze which explanation best fits all observed symptoms
4. Recommend specific diagnostic tests to confirm your hypothesis
5. Suggest immediate mitigation actions while waiting for test results

Applied Example:

```
Think step-by-step to diagnose the following problem:
The atmospheric distillation tower shows a gradual increase in top
section temperature (from 120 to 135°C) over the past 72 hours, while:
```

- Crude flow has remained constant at 120,000 BPD
- Furnace outlet temperature has stayed at 360 °C ± 3 °C
- Overhead accumulator pressure has increased from 1.2 to 1.5 kg/cm^2
- Heavy naphtha end point has increased from 165 to 175 °C
- Circulating reflux shows gradual darkening

1. First, identify potential causes that could explain these symptoms
2. For each potential cause, evaluate the evidence for and against
3. Analyze which explanation best fits all observed symptoms

4. Recommend specific diagnostic tests to confirm your hypothesis
5. Suggest immediate mitigation actions while waiting for test results.

Applied Technique: Chain of Thought (Sect. 4.2, "Second Generation Techniques"). This technique guides systematic reasoning through complex diagnostic problems by explicit step-by-step analysis.

B.2.2 Zero-Shot Chain of Thought: Catalyst Selection

General Template:

```
Let's think step-by-step to [objective] based on the following informa-
tion:
[Detailed contextual information]
Considering each factor systematically:
```

Applied Example:

```
Let's think step-by-step to select the optimal catalyst for a diesel
hydrotreater revamp based on the following information:
Unit parameters:
```

- Feed capacity: 25,000 BPD
- Feed properties: 1500 ppm sulfur, 250 ppm nitrogen, 25% LCO content
- Product specs: < 10 ppm sulfur, < 10 ppm nitrogen, Cetane index > 51
- Existing reactor volume: 85 m^3
- Operating pressure limitation: 65 barg maximum
- Available hydrogen: 150 Nm^3/m^3 feed
- Expected cycle length: 24-36 months.

```
Candidate catalysts:
```

```
A. Conventional NiMo on alumina (high activity, moderate metals tolerance) B. Advanced NiMo with modified support (very high activity, moderate metals tolerance) C. CoMo with Type II active sites (moderate activity, high selectivity, excellent metals tolerance) D. Graded system with specialized top layer (high activity, excellent metals tolerance, higher pressure drop).
Considering each factor systematically:
```

Applied Technique: Zero-Shot Chain of Thought (Sect. 4.2.2 "Second Generation Techniques"). This technique triggers structured reasoning without requiring explicit examples, using simple prompting to encourage step-by-step analysis.

B.2.3 Role Prompting: Energy Efficiency Audit

General Template:

```
As a [specialist type] with [years] of experience specializing in [specific domain expertise], analyze the following situation:
[Detailed description]
Based on your expertise:
```

1. `[Required analysis component 1]`
2. `[Required analysis component 2]`
3. `[Required analysis component 3]`

Applied Example:

```
As a senior energy optimization specialist with 25 years of experience in refinery heat integration and process efficiency, analyze the following situation:
A 150,000 BPD crude distillation unit shows the following energy parameters:
```

- `Current energy intensity: 1.35 MBtu/bbl (industry best quartile: 0.9 MBtu/bbl)`
- `Furnace efficiency: 82% (stack temperature 230 °C)`
- `Preheat train temperature: crude inlet 20 °C, furnace inlet 235 °C`

- Heat exchanger network: 24 exchangers in series/parallel configuration
- Steam generation: 45 tonnes/hr at 45 barg, 15 tonnes/hr at 10 barg
- Pumparound duties: top 25 MW, upper 30 MW, middle 40 MW, bottom 35 MW
- Current pinch temperature: approximately 260 °C
- Minimal heat integration with adjacent units

Based on your expertise:

1. Identify the three highest-impact opportunities for energy efficiency improvement
2. For each opportunity, provide specific technical recommendations with expected energy savings
3. Outline implementation requirements including capital investment, complexity, and downtime needs
4. Recommend a phased implementation approach with prioritization rationale
5. Define key performance indicators to track improvement effectiveness

Applied Technique: Role Prompting (Sect. 4.2.2 "Second Generation Techniques"). This technique directs the LLM to adopt a specific professional perspective, activating domain-specific knowledge patterns for specialized analysis.

B.3 Third-Generation Techniques

B.3.1 Tree of Thoughts: Hydrogen Network Optimization

General Template:

```
Explore multiple solution pathways for [objective] using a tree of thoughts approach:
[Problem description with constraints]
Branch 1: [First major decision point]
```

- Option 1A: [Alternative 1 for Branch 1]
- Option 1B: [Alternative 2 for Branch 1]
- Option 1C: [Alternative 3 for Branch 1]

Branch 2: [Second major decision point]

- Option 2A: [Alternative 1 for Branch 2]
- Option 2B: [Alternative 2 for Branch 2]
- Option 2C: [Alternative 3 for Branch 2]

Branch 3: [Third major decision point]

- Option 3A: [Alternative 1 for Branch 3]
- Option 3B: [Alternative 2 for Branch 3]
- Option 3C: [Alternative 3 for Branch 3]

For each branch, analyze the implications, trade-offs, and interdependencies. Then identify the most promising combination of options across all branches.

Applied Example:

Explore multiple solution pathways for hydrogen network optimization using a tree of thoughts approach:
A refinery hydrogen system has the following characteristics:

- Current hydrogen production: 80 MMSCFD from steam methane reformer (99.9% purity)
- Catalytic reformer hydrogen: 15 MMSCFD (80% purity)
- Consumers: hydrocracker (45 MMSCFD, 99 + % purity), diesel hydrotreater (30 MMSCFD, 90 + % purity), naphtha hydrotreater (15 MMSCFD, 85 + % purity)
- Current purification: Single PSA unit (50 MMSCFD capacity)
- Pressure levels: high (150 bar), medium (80 bar), low (30 bar)

- Average hydrogen losses: 15% through purging and inefficient use
- Natural gas price: $6/MMBtu, Carbon price: $40/tonne CO_2

Branch 1: Production strategy

- Option 1A: Expand existing SMR capacity
- Option 1B: Install electrolysis capacity
- Option 1C: Optimize existing units without capacity addition

Branch 2: Recovery and purification approach

- Option 2A: Additional PSA capacity
- Option 2B: Membrane separation technology
- Option 2C: Cryogenic purification for high-purity needs

Branch 3: Network configuration

- Option 3A: Centralized management system
- Option 3B: Segregated networks for different purity levels
- Option 3C: Cascade system with quality-appropriate routing

For each branch, analyze the implications, trade-offs, and interdependencies. Then identify the most promising combination of options across all branches.

Applied Technique: Tree of Thoughts (Sect. 4.2.3, "Third Generation Techniques"). This technique creates a branching structure to explore multiple solution pathways simultaneously, comparing their implications for complex decisions.

B.3.2 ReAct: Process Safety Hazard Analysis

General Template:

```
Using the ReAct (Reasoning and Acting) approach, develop a systematic
analysis for [objective]:
Thought: [Analyze what you know about the situation and what information
you need]
Action: [Determine the most appropriate next step based on your thinking]
Observation: [Note what you learn from the action]
Thought: [Analyze this new information and determine next steps]
Action: [Specify next analysis or investigation]
...
Continue this reasoning and action cycle until you have a complete anal-
ysis with decision points for different scenarios.
```

Applied Example:

```
Using the ReAct (Reasoning and Acting) approach, develop a systematic
process hazard analysis for a hydrocracker depressuring system:
Thought: I need to understand the hydrocracker depressuring system's
purpose and critical components. This system is a safety-critical emer-
gency system that rapidly reduces pressure in the reactor section during
abnormal conditions.
Action: Identify the key components and failure modes of a typical hydro-
cracker depressuring system.
Observation: Key components include depressuring valves (typically fail-
open), blowdown header, knockout drum, quench system, and flare system.
Critical failure modes include valve failure to open, header blockage,
insufficient header/flare capacity, and control system failure.
Thought: Now I should analyze the potential hazard scenarios that would
require depressuring system activation.
Action: Enumerate the primary scenarios requiring emergency depressuring
and their initiating events.
Observation: Primary scenarios include: 1) Loss of recycle gas compres-
sor, 2) Exothermic runaway reaction, 3) Major leak in high-pressure sys-
tem, 4) Loss of reactor quench, 5) Fire in the unit affecting pressure-
containing equipment.
Thought: I need to evaluate the potential consequences if the depressur-
ing system fails to operate correctly in each scenario.
Action: Analyze consequence severity for each scenario with depressuring
system failure.
...
```

```
Continue this reasoning and action cycle until you have a complete hazard
analysis with decision points for different scenarios.
```

Applied Technique: ReAct (Sect. 4.2.3, "Third Generation Techniques"). This technique integrates structured reasoning with planned action steps, enabling complex problem solving through an iterative cycle of thought and action.

B.4 Integrated Approaches

B.4.1 Retrieval Augmented Generation: Turnaround Planning

General Template:

```
Based on the following reference information and industry best practices,
develop a [specific deliverable] for [objective]:
Reference data:
[Relevant technical specifications, historical data, or reference infor-
mation]
Project parameters:
[Specific constraints, requirements, and objectives]
Develop a comprehensive plan that:
```

1. [Required component 1]
2. [Required component 2]
3. [Required component 3]

Applied Example:

```
Based on the following reference information and industry best prac-
tices, develop a critical path schedule for a fluid catalytic cracking
unit turnaround:
Reference data:
```

- Previous turnaround duration: 28 days
- Critical path in previous turnaround: regenerator refractory replacement (14 days)
- Historical performance: reactor internal replacement (10 days), main air blower overhaul (12 days), expansion joint replacements (6 days)

- Industry benchmarks: comparable FCCUs complete turnarounds in 21–24 days
- Mechanical availability of specialized crews: refractory (limited to 3 crews), catalyst handling (2 crews), vessel entry (4 crews)

Project parameters:

- Scope includes: complete reactor internal replacement, 60% regenerator refractory replacement, full turboexpander overhaul, 4 major heat exchangers, relief valve servicing, regenerator cyclone replacement
- Schedule constraints: must complete within 24 days maximum
- Resource constraints: limited availability of specialized refractory installers
- Regulatory inspections required for 12 pressure vessels
- Available workforce: 350 craft workers maximum

Develop a comprehensive plan that:

1. Identifies the critical path and near-critical activities
2. Proposes specific schedule optimization techniques
3. Addresses resource constraints with mitigation strategies
4. Includes contingency planning for high-risk activities
5. Provides key milestones and decision points for management

Applied Technique: Retrieval Augmented Generation (Sect. 4.2.4, "Integrated Approaches"). This technique combines LLM capabilities with specific contextual information, enabling more precise and relevant outputs for specialized applications.

B.4.2 Multi-perspective Analysis: Carbon Reduction Strategy

General Template:

Analyze [topic/objective] from the following perspectives:

1. [Perspective 1]: [Brief description of viewpoint]
2. [Perspective 2]: [Brief description of viewpoint]
3. [Perspective 3]: [Brief description of viewpoint] ...

For each perspective, consider:

- Primary concerns and priorities
- Key success metrics
- Potential challenges and limitations
- Specific recommendations

Then provide an integrated analysis that balances these perspectives into a coherent strategy.

Applied Example:

Analyze a refinery carbon reduction strategy from the following perspectives:

1. Technical Operations: Focus on process efficiency, energy integration, and technology feasibility
2. Economic Analysis: Focus on capital requirements, operating costs, and return on investment
3. Regulatory Compliance: Focus on emissions standards, carbon pricing, and reporting requirements
4. Strategic Planning: Focus on long-term positioning, market trends, and competitive advantage
5. Stakeholder Relations: Focus on community impact, investor expectations, and public perception

For each perspective, consider:

- Primary concerns and priorities
- Key success metrics
- Potential challenges and limitations

- Specific recommendations

Then provide an integrated analysis that balances these perspectives into a coherent carbon reduction strategy for a medium-sized refining company.

Applied Technique: Multi-perspective Analysis (Sect. 4.2.4, "Multi-perspective Analysis"). This technique examines challenges from diverse viewpoints simultaneously to ensure comprehensive understanding before developing solutions.

B.4.3 First Principles with Visualization Prompting: Heat Integration

General Template:

Using first principles of [domain science], analyze [technical objective].
[Detailed parameter specifications]

1. Begin by establishing the fundamental laws and equations governing this system
2. Perform step-by-step analysis based on these principles
3. Create a clear visualization concept that illustrates:
   - [Visualization element 1]
   - [Visualization element 2]
   - [Visualization element 3]

Provide both the technical analysis and a detailed description of how this should be visualized for maximum clarity.

Applied Example:

Using first principles of thermodynamics and heat transfer, analyze heat integration opportunities between a crude distillation unit and hydrocracker.
System parameters:

- Crude unit (150,000 BPD):

  - Atmospheric overhead: 160 °C → 45 °C, 4.5 MW
  - Kerosene pumparound: 210 °C → 175 °C, 3.0 MW
  - Diesel pumparound: 290 °C → 240 °C, 4.0 MW
  - AGO pumparound: 340 °C → 270 °C, 3.5 MW
- Hydrocracker (50,000 BPD):
  - Reactor effluent: 410 °C → 230 °C, 6.5 MW
  - Fractionator overhead: 150 °C → 40 °C, 2.5 MW
  - Feed preheat requirement: 80 °C → 380 °C, 7.5 MW

1. Begin by establishing the fundamental laws and equations governing this system
2. Perform step-by-step analysis based on these principles
3. Create a clear visualization concept that illustrates:
   - Temperature-enthalpy profiles for hot and cold streams
   - Composite curves showing integration opportunities
   - Proposed heat exchanger network configuration
   - Current vs. optimized energy consumption

Provide both the technical analysis and a detailed description of how this should be visualized for maximum clarity.

Applied Technique: First Principles Reasoning with Visualization Prompting (Sect. 4.2.4 "Visualization Prompting"). This technique combines scientific fundamentals with guidance to develop visual representations of complex systems.

B.5 Implementation-Focused Approaches

B.5.1 Risk-Opportunity Mapping: Digital Transformation

General Template:

Develop a risk-opportunity map for implementing [technology/system] in [industry context], analyzing both challenges requiring mitigation and potential advantages.
Systematically identify and analyze these risk-opportunity pairs:
Technical domain:

- Risk: [Technical risk description]

- Potential impact, probability, early indicators
- Mitigation strategies and contingency options
- Opportunity: [Technical opportunity description]
  - Potential benefits, likelihood, triggering events
  - Approaches to position for capture of this opportunity

[Additional domains with risks and opportunities]
For each risk-opportunity pair, provide:

1. Detailed impact analysis (quantified where possible)
2. Probability assessment with supporting rationale
3. Key indicators or triggers to monitor
4. Specific action plan with both mitigation and opportunity capture

Applied Example:

Develop a risk-opportunity map for implementing advanced process control (APC) and real-time optimization (RTO) in a complex refinery, analyzing both challenges requiring mitigation and potential advantages.
Systematically identify and analyze these risk-opportunity pairs:
Technical domain:

- Risk: Model degradation and maintenance challenges
  - Potential impact, probability, early indicators
  - Mitigation strategies and contingency options
- Opportunity: Integration with digital twin technology
  - Potential benefits, likelihood, triggering events
  - Approaches to position for capture of this opportunity

Operational domain:

- Risk: Operator resistance and overreliance on automation
  - Potential impact, probability, early indicators
  - Mitigation strategies and contingency options
- Opportunity: Knowledge capture from experienced operators
  - Potential benefits, likelihood, triggering events

  - Approaches to position for capture of this opportunity

Economic domain:

- Risk: Difficult ROI quantification and delayed benefits
  - Potential impact, probability, early indicators
  - Mitigation strategies and contingency options
- Opportunity: Performance-based vendor contracts
  - Potential benefits, likelihood, triggering events
  - Approaches to position for capture of this opportunity

For each risk-opportunity pair, provide:

1. Detailed impact analysis (quantified where possible)
2. Probability assessment with supporting rationale
3. Key indicators or triggers to monitor
4. Specific action plan with both mitigation and opportunity capture

Applied Technique: Risk-Opportunity Mapping (Sect. 5.7.1 "Risk Opportunity Mapping"). This technique systematically identifies both challenges requiring mitigation and potential advantages that could be leveraged during implementation.

B.5.2 Phased Implementation Strategy: Renewable Integration

General Template:

```
Develop a phased implementation strategy for [objective] that manages
risks while enabling progressive capability building:
Current state assessment:
[Detailed description of current situation and baseline]
Target end state:
[Description of desired future state and key success metrics]
Define implementation phases:
Phase 1 (Months X-Y): [Phase name - Foundation]
```

- Key activities and deliverables
- Required resources and investments
- Expected outcomes and transition criteria

[Additional phases with similar structure]
For the overall implementation:

1. Define critical success factors and potential showstoppers
2. Establish governance structure and decision points
3. Outline key performance indicators for tracking progress
4. Identify contingency plans for major risk areas

Applied Example:

Develop a phased implementation strategy for integrating renewable feedstocks into an existing refinery that manages risks while enabling progressive capability building:
Current state assessment:

- 250,000 BPD conventional petroleum refinery with hydroprocessing, FCC, reforming, and coking
- No current renewable processing capability
- Available plot space adjacent to diesel hydrotreater
- Limited experience with bio-feedstocks
- Regional regulations requiring 10% renewable content within 3 years, 20% within 5 years

Target end state:

- 30,000 BPD renewable processing capacity (12% of total throughput)
- Flexibility to process multiple feedstock types (vegetable oils, waste fats, pyrolysis oils)
- Full regulatory compliance with product certification
- Carbon intensity reduction of 15-20% across product portfolio

- Maintenance of product quality specifications and reliability metrics

Define implementation phases:
Phase 1 (Months 1-6): Foundation Building

- Detailed feedstock market assessment and supply chain development
- Laboratory and pilot testing of candidate feedstocks
- Engineering studies and technology selection
- Initial procurement of long-lead equipment
- Staff training program development
- Expected outcome: Validated technical approach and baseline performance data

Phase 2 (Months 7-18): Initial Implementation

- Installation of dedicated pre-treatment unit (5000 BPD)
- Modifications to existing hydrotreater for co-processing (5-10% renewable)
- Control system updates and laboratory capability enhancement
- Development of quality control protocols and certification procedures
- Expected outcome: Initial renewable diesel production at 5-7% of diesel output

Phase 3 (Months 19-36): Capability Expansion

- Installation of dedicated renewable processing train (15,000 BPD)
- Hydrogen system upgrades to support increased demand
- Product certification and marketing development
- Supply chain optimization and diversification

- Expected outcome: Renewable processing at 8–10% of total throughput

Phase 4 (Months 37–48): Full Integration

- Second-phase capacity addition (10,000 BPD)
- Advanced process control implementation
- Full integration with existing production planning systems
- Development of proprietary technology improvements
- Expected outcome: Target capacity of 30,000 BPD with full flexibility

For the overall implementation:

1. Define critical success factors and potential showstoppers
2. Establish governance structure and decision points
3. Outline key performance indicators for tracking progress
4. Identify contingency plans for major risk areas

Applied Technique: Phased Implementation Strategy (Sect. 5.7.2 "Implementation Planning Phase"). This technique develops staged approaches that manage risk while enabling progressive learning and capability building.

Appendix C: Performance Data, Economic Metrics, and Environmental Impact for Net Zero Refinery Design

C.1 Introduction

This appendix provides comprehensive data sets that informed the net zero refinery design presented in Chap. 5. The following sections present performance metrics, capital and operating expenditures, energy consumption patterns, carbon footprint analyses, and technical parameters. Each table includes relevant bibliographic citations to support the figures presented. These data points formed the foundation for design decisions, technology selection, and economic modeling throughout the design process.

C.2 Feedstock Characteristics and Availability

C.2.1 Municipal Solid Waste (MSW)

See Table C.1.

Table C.1 MSW feedstock characteristics

Parameter	Value	Source
Average calorific value	10–13 MJ/kg	Arena and Di Gregorio (2014)
Carbon content	35–45% by weight	Zhou et al. (2015)
Biogenic carbon fraction	50–65%	Obermeier et al. (2021)
Sorting efficiency (advanced)	85–95%	Gundupalli et al. (2017)
Collection radius economics	30–50 km	Merrild et al. (2012)
Seasonality variation	± 12%	Denafas et al. (2014)
Fee range	$25–85/tonne	Hoornweg and Bhada-Tata (2012)

R. Larraz, *Prompt Engineering and the Transformation of Petroleum Refining*, Synthesis Lectures on Chemical Engineering and Biochemical Engineering,
https://doi.org/10.1007/978-3-031-99728-0

C.2.2 Agricultural and Forestry Residues

See Table C.2.

Table C.2 Agricultural and forestry residue characteristics

Parameter	Value	Source
Agricultural residue yield	0.8–1.7 dry tonnes/ha	Bentsen et al. (2014)
Forestry residue availability	20–35% of harvested wood	Egnell and Björheden (2012)
Average moisture content	30–55%	Vassilev et al. (2010)
Carbon content (dry basis)	40–50%	Vassilev et al. (2010)
Collection cost range	$40–85/dry tonne	FAO 2016
Sustainable removal rate	35–50% of residues	Bronick et al. (2005)
Energy density (dried)	15–19 MJ/kg	McKendry (2002)
Seasonal availability variation	±30–45%	Rentizelas et al. (2009)

C.2.3 Waste Plastics

See Table C.3.

Table C.3 Waste plastic feedstock characteristics

Parameter	Value	Source
Calorific value	35–45 MJ/kg	Al-Salem et al. (2017)
Carbon content	75–90% by weight	Sharuddin et al. (2016)
Collection cost	$150–400/tonne	Hopewell et al. (2009)
Contamination level	5–15% by weight	Ragaert et al. (2017)
Polyolefin content (PE, PP)	55–70% of waste plastics	Plastics Europe (2019)
PET/PVC/PS content	20–35% of waste plastics	Geyer et al. (2017)
Global recycling rate	9–14%	Ritchie and Roser (2018)
Sorting efficiency	80–95% (advanced)	Gundupalli et al. (2017)

C.3 Technology Performance Data

C.3.1 Waste Sorting and Preprocessing

See Table C.4.

Table C.4 Waste sorting and preprocessing performance metrics

Parameter	Value	Source
Advanced NIR sorting accuracy	92–98%	Gundupalli et al. (2017)
Energy consumption for sorting	15–45 kWh/tonne	Cimpan et al. (2015)
Processing capacity	10–50 tonnes/hr	Bonifazi and Serranti (2019)
Capital cost	$10–25 million (100,000 tpy)	Pires et al. (2011)
Operating cost	$15–40/tonne	Cimpan et al. (2016)
Material recovery efficiency	80–95%	Todd et al. (2015)
Economic throughput minimum	30,000 tonnes/year	Pressley et al. (2014)
System availability	85–92%	Author estimation

C.3.2 Catalytic Pyrolysis

See Table C.5.

Table C.5 Catalytic pyrolysis performance data

Parameter	Value	Source
Liquid yield (plastics)	70–85 wt%	Miandad et al. (2016)
Liquid yield (biomass)	50–65 wt%	Bridgwater (2012)
Gas yield	15–25 wt%	Wang et al. (2017)
Catalyst lifetime	200–1000 h	Iliopoulou et al. (2012)
Energy requirement	1.5–2.0 MJ/kg feed	Abnisa and Wan Daud (2014)
Carbon efficiency	65–80%	Liu et al. (2014)
Capital cost	$15–25 million (100 tpd)	Author estimation
Operating cost	$50–90/tonne input	Wright et al. (2010)
Bio-oil oxygen content	20–40 wt%	Venderbosch and Prins (2010)
Plastic oil aromatic content	15–40 wt%	Williams (2013)

C.3.3 Oxygen-Blown Biomass Gasification

See Table C.6.

Table C.6 Biomass gasification performance metrics

Parameter	Value	Source
Carbon conversion	85–95%	Prins et al. (2007)
Cold gas efficiency	70–80%	Molino et al. (2016)
H_2/CO ratio range	0.8–2.0	Sikarwar et al. (2016)
Tar content in raw gas	5–50 g/Nm^3	Milne et al. (1998)
Particulate loading	10–100 g/Nm^3	van der Drift et al. (2004)
Energy consumption	0.1–0.3 MWh/tonne	Arena (2012)
Capital cost	\$50–85 million (800 tpd)	Swanson et al. (2010)
Operating cost	\$35–65/tonne biomass	Dutta et al. (2015)
System availability	75–85%	Kurkela et al. (2016)
Scale-up factor	0.7–0.8	Larson et al. (2009)

C.3.4 Fischer–Tropsch Synthesis

See Table C.7.

Table C.7 Fischer–Tropsch synthesis performance data

Parameter	Value	Source
Carbon conversion efficiency	60–75%	Dry (2002)
Selectivity to C_5+	70–85%	Davis (2001)
Cobalt catalyst lifetime	3–5 years	Ail and Dasappa (2016)
Iron catalyst lifetime	8–12 months	de Smit and Weckhuysen (2008)
H_2/CO optimal ratio	1.8–2.1 (cobalt); 0.7–1.4 (iron)	Jahangiri et al. (2014)
Energy release	165–170 kJ/mol CO	Vervloet et al. (2012)
Temperature range	200–250 °C (cobalt); 320–350 °C (iron)	Schulz (1999)
Pressure range	15–40 bar	van der Laan and Beenackers (1999)
Capital cost	\$40–60 million (1000 bpd)	Swanson et al. (2010)

(continued)

(continued)

Parameter	Value	Source
Operating cost	\$8–14/GJ product	Bechtel (1998) updated to present time

C.4 Economic Metrics

C.4.1 Capital Expenditure (CAPEX) Breakdown

See Table C.8.

Table C.8 Capital expenditure breakdown by process unit

Process unit	Capacity	CAPEX range (\$million)	Reference source
Waste sorting and preprocessing	300,000 tpy	\$45–75	Cimpan et al. (2016)
Biomass preprocessing	600,000 tpy	\$80–120	Searcy and Hess (2010)
Plastics pyrolysis	150,000 tpy	\$90–140	Fivga and Dimitriou (2018)
Biomass gasification	800 tpd	\$250–400	Swanson et al. (2010)
Syngas cleaning	150,000 Nm^3/hr	\$70–120	Dutta et al. (2015)
Fischer–Tropsch synthesis	2000 bpd	\$180–280	Bechtel (1998), updated
Alcohol synthesis	500 tpd	\$70–110	Phillips et al. (2007)
Hydroprocessing	10,000 bpd	\$130–210	Author estimation
CO_2 capture	500,000 tpy	\$70–120	Rubin et al. (2015)
CO_2utilization	100,000 tpy	\$80–150	Pérez-Fortes et al. (2016)
Hydrogen production	30 tpd	\$100–180	IRENA (2020)
Utilities and infrastructure	–	\$300–450	Estimated from Peters et al. (2003)
Engineering and contingency	–	\$350–550	25–30% of direct costs
Total CAPEX	–	\$3.8–4.2 billion	Compiled from above

Note All figures adjusted to 2025 dollars using Chemical Engineering Plant Cost Index (CEPCI) projections

C.4.2 Operating Expenditure (OPEX) Breakdown

See Table C.9.

Table C.9 Annual operating expenditure breakdown

Category	Annual cost ($million)	Basis	Source
Feedstock costs (net)	$150–200	Mixed portfolio with tipping fees offsetting costs	Compiled from feedstock data
Electricity	$40–70	500–700 GWh at $80–100/MWh	IEA (2020)
Natural gas (auxiliary)	$20–35	300–500 TJ at $6–7/GJ	EIA (2020)
Catalysts and chemicals	$20–35	Based on consumption rates	Estimated from technical data
Labor (200–250 staff)	$25–35	$125,000–140,000 per employee	Industry benchmarks
Maintenance	$50–65	2.5–3% of CAPEX annually	Peters et al. (2003)
Insurance and property tax	$15–25	0.5–0.7% of CAPEX annually	Industry benchmarks
Other overhead	$30–50	Administration, marketing, etc.	Industry benchmarks
Total OPEX	$350–515 million	–	Compiled from above

C.4.3 Revenue Structure

See Table C.10.

Table C.10 Annual revenue structure by product

Product	Annual production	Price range	Annual revenue ($million)	Source
Sustainable aviation fuel	600,000 bbl	$150–190/bbl	$90–115	ICAO
Renewable diesel	1,500,000 bbl	$135–165/bbl	$200–250	Argus Media
Renewable gasoline	800,000 bbl	$125–155/bbl	$100–125	Calculated from market premiums

(continued)

(continued)

Product	Annual production	Price range	Annual revenue ($million)	Source
Biochemicals	150,000 tonnes	\$1200–1800/ tonne	\$180–270	ICIS
Bio-naphtha	200,000 tonnes	\$900–1100/ tonne	\$180–220	Platts
Carbon credits	500,000 tonnes CO_2e	\$50–120/tonne	\$25–60	Projected from IEA (2021)
Renewable energy credits	Various	–	\$15–35	Based on regional programs
Total revenue	–	–	\$790–1075 million	Compiled from above

C.4.4 Financial Performance Metrics

See Table C.11.

Table C.11 Key financial performance indicators

Metric	Value	Basis	Source
Internal rate of return (IRR)	10–14%	Base case scenario	Financial model
Net present value (NPV)	\$0.9–1.4 billion	At 8% discount rate	Financial model
Payback period	7–9 years	Simple payback	Financial model
EBITDA margin	32–38%	Based on revenue/ OPEX	Financial model
Capital intensity	\$38,000–42,000/bpd equivalent	Total CAPEX per barrel capacity	Calculated
Carbon abatement cost	\$30–60/tonne CO_2e	Net cost per tonne of emissions avoided	Calculated
Minimum sustainable aviation fuel premium	15–25%	Premium over conventional required for viability	Sensitivity analysis

C.5 Energy Consumption and Efficiency

C.5.1 Energy Consumption by Process Area

See Table C.12.

Table C.12 Energy consumption breakdown by process area

Process area	Electricity (GWh/yr)	Thermal energy (TJ/yr)	Source
Waste sorting and preprocessing	35–50	100–150	Cimpan et al. (2015)
Biomass preprocessing	80–120	650–850	Miao et al. (2011)
Plastics pyrolysis	30–45	250–350	Abnisa and Wan Daud (2014)
Biomass gasification	70–100	500–700	Molino et al. (2016)
Syngas processing	60–90	150–250	Dutta et al. (2015)
Fischer–Tropsch synthesis	40–60	− 400 to − 600 (exothermic)	Vervloet et al. (2012)
Hydroprocessing	50–75	300–450	Author estimation
CO_2 capture and utilization	100–150	450–650	Rubin et al. (2015)
Hydrogen production	150–200	10–30	Carmo et al. (2013)
Utilities and auxiliary	30–50	100–200	Estimated
Total consumption	645–940 GWh/yr	2110–3030 TJ/yr	Compiled

Note Negative thermal energy indicates net heat production (exothermic processes)

C.5.2 Energy Efficiency Metrics

See Table C.13.

Table C.13 Overall energy efficiency indicators

Metric	Value (%)	Basis	Source
Overall energy efficiency	42–55	Energy in products/Energy in feedstock + utilities	Calculated
Thermal integration efficiency	65–75	Heat recovered/Total heat available	Heat integration analysis
Biomass-to-liquids efficiency	40–48	Energy in liquid products/Energy in biomass	Calculated from Swanson et al. (2010)
Waste-to-energy efficiency	55–65	Energy recovered/Energy in waste	Calculated
Plastics-to-fuel efficiency	75–85	Energy in fuel/Energy in plastic	Calculated from Miandad et al. (2016)
Hydrogen utilization efficiency	80–90	H_2 effectively used/H_2 produced	Process modeling
CO_2 utilization energy efficiency	35–45	Energy in products/Energy input for CO_2 conversion	Process modeling

C.6 Carbon Footprint and Environmental Impact

C.6.1 Carbon Intensity by Process Stage

See Table C.14.

Table C.14 Carbon intensity breakdown by process stage

Process stage	Carbon intensity (kg CO_2e/GJ product)	Basis	Source
Feedstock collection and transport	3–8	Diesel usage, transportation distances	Edwards et al. (2014)
Preprocessing and sorting	5–10	Electricity and thermal energy	Cimpan et al. (2016)
Thermochemical conversion	8–15	Process emissions, energy use	Swanson et al. (2010)

(continued)

(continued)

Process stage	Carbon intensity (kg CO_2e/GJ product)	Basis	Source
Upgrading and refining	4–9	Hydrogen usage, process energy	Ail and Dasappa (2016)
Utilities and auxiliaries	2–5	Support systems energy use	Estimated
Carbon capture credit	− 40 to − 60	Biogenic carbon sequestration	Process modeling
Net carbon intensity	− 18 to − 28	Full lifecycle basis	Calculated

Note Negative values indicate net carbon removal from atmosphere over lifecycle

C.6.2 Carbon Reduction Compared to Conventional Refining

See Table C.15.

Table C.15 Carbon intensity comparison with conventional products

Product	Conventional carbon intensity (g CO_2e/MJ)	Net zero design carbon intensity (g CO_2e/MJ)	Reduction (%)	Source
Jet fuel	85–95	− 5 to − 15	>100%	Edwards et al. (2014), Design calculation
Diesel	80–90	− 5 to − 15	>100%	Edwards et al. (2014), Design calculation
Gasoline	85–95	− 5 to − 15	>100%	Edwards et al. (2014), Design calculation
Naphtha	70–80	− 5 to − 15	>100%	Edwards et al. (2014), Design calculation

C.7 Implementation Timeline and Phasing

C.7.1 Implementation Schedule

See Table C.16.

Table C.16 Implementation phases and timeline

Phase	Timeframe	Key activities	CAPEX allocation	Source
Foundation phase	Years 1–3	Waste preprocessing, Initial pyrolysis, first gasification train	\$1.2–1.5 billion	Implementation plan
Expansion phase	Years 4–6	Additional processing capacity, advanced catalysts, Initial carbon capture	\$1.1–1.4 billion	Implementation plan
Integration phase	Years 7–10	Full heat integration, complete carbon management, Renewable energy portfolio	\$0.9–1.2 billion	Implementation plan
Evolution phase	Years 11+	Technology upgrades, capacity optimization, Potential diversification	\$0.5–0.8 billion	Implementation plan

C.7.2 Technology Readiness Levels and Implementation Risk

See Table C.17.

Table C.17 Technology readiness assessment

Technology	Current TRL	Implementation risk level	Key development needs	Source
Advanced waste sorting	8–9	Low	Process optimization	Gundupalli et al. (2017)
Catalytic pyrolysis	6–7	Medium–high	Scale-up, catalyst stability	Miandad et al. (2016)
Biomass gasification	7–8	Medium	Reliability, tar management	Molino et al. (2016)
Hot gas cleanup	6–7	Medium–high	Catalyst poisoning resistance	Woolcock and Brown (2013)

(continued)

(continued)

Technology	Current TRL	Implementation risk level	Key development needs	Source
Fischer–Tropsch synthesis	8–9	Low	Optimization for biomass syngas	Ail and Dasappa (2016)
CO_2 capture	7–8	Medium	Energy efficiency	Rubin et al. (2015)
CO_2 utilization	5–6	High	Catalyst development, economics	Porosoff et al. (2016)
Green hydrogen production	7–8	Medium	Scale-up, cost reduction	IRENA (2020)
Renewable energy integration	7–8	Medium	Storage optimization	Pantaleo et al. (2021)

References

Abnisa F, Wan Daud WMA (2014) A review on co-pyrolysis of biomass: an optional technique to obtain a high-grade pyrolysis oil. Energy Convers Manage 87: 71–85

Ail SS, Dasappa S (2016) Biomass to liquid transportation fuel via Fischer Tropsch synthesis–technology review and current scenario. Renew Sustain Energy Rev 58:267–286

Al-Salem SM, Antelava A, Constantinou A, Manos G, Dutta A (2017) A review on thermal and catalytic pyrolysis of plastic solid waste (PSW). J Environ Manage 197:177–198

Arena U (2012) Process and technological aspects of municipal solid waste gasification. A review. Waste Manag 32(4):625–639

Arena U, di Gregorio F (2014) Energy generation by air gasification of two industrial plastic wastes in a pilot scale fluidized bed reactor. Energy 68:735–743

Bechtel (1998) Baseline Design/Economics for advanced Fischer-Tropsch technology. Final report—April, 1998

Bentsen NS, Felby C, Thorsen BJ (2014) Agricultural residue production and potentials for energy and materials services. Prog Energy Combust Sci 40:59–73

Bonifazi G, Serranti S (2019) Techniques for separation of plastic wastes. Use of recycled plastics in eco-efficient concrete. Woodhead Publishing, pp 9–37

Bridgwater AV (2012) Review of fast pyrolysis of biomass and product upgrading. Biomass Bioenerg 38:68–94

Bronick CJ, Lal R (2005) Soil structure and management: a review. Geoderma 124(1–2):3–22

Carmo M, Fritz DL, Mergel J, Stolten D (2013) A comprehensive review on PEM water electrolysis. Int J Hydrogen Energy 38(12):4901–4934

Cimpan C, Maul A, Jansen M, Pretz T, Wenzel H (2015) Central sorting and recovery of MSW recyclable materials: a review of technological state-of-the-art, cases, practice and implications for materials recycling. J Environ Manage 156:181–199

Cimpan C, Maul A, Pretz T, Wenzel H (2016) Techno-economic assessment of central sorting at material recovery facilities—the case of lightweight packaging waste. J Clean Prod 112(5):4837–4397

Davis BH (2001) Fischer-Tropsch synthesis: current mechanism and futuristic needs. Fuel Process Technol 71(1–3):157–166

Denafas G, Ruzgas T, Martuzevičius T, Shmarin S, Hoffmann M, Mykhaylenko V, Ogorodnik S, Romanov M, Neguliaeva E, Chusov A, Turkadze T, Bochoidze I, Ludwig C (2014) Seasonal variation of municipal solid waste generation and composition in four East European cities. Resour, Conserv Recycl 89:22–30

Dry ME (2002) The Fischer-Tropsch process: 1950–2000. Catal Today 71(3–4):227–241

Dutta A, Sahir A, Tan E, Humbird D, Snowden-Swan LJ, Meyer P, Ross J, Sexton D, Yap R, Lukas J (2015) Process design and economics for the conversion of lignocellulosic biomass to hydrocarbon fuels. National Renewable Energy Laboratory, Report No. NREL/TP-5100-62455

Edwards R, Larivé JF, Rickeard D, Weindorf W (2014) Well-to-wheels analysis of future automotive fuels and powertrains in the European context. JRC Technical Reports, Report EUR, 26236

Egnell G, Björheden R (2012) Options for increasing biomass output from long-rotation forestry. Wires Energy Environ 2(4):465–472

EIA (2020) Annual energy outlook 2020. U.S. Energy Information Administration, Washington, DC

FAO (2016) Handbook on agricultural cost of production statistics

Fivga A, Dimitriou I (2018) Pyrolysis of plastic waste for production of heavy fuel substitute: a techno-economic assessment. Energy 149:865–874

Geyer R, Jambeck JR, Law KL (2017) Production, use, and fate of all plastics ever made. Sci Adv 3(7)

Gundupalli SP, Hait S, Thakur A (2017) A review on automated sorting of source-separated municipal solid waste for recycling. Waste Manage 60:56–74

Hoornweg D, Bhada-Tata P (2012) What a waste: a global review of solid waste management. World Bank Urban Dev Ser Knowl Pap 15:1–98

Hopewell J, Dvorak R, Kosior E (2009) Plastics recycling: challenges and opportunities. Philos Trans R Soc Lond B Biol Sci 27, 364(1526):2115–2126

IEA (2020) Global energy review 2020. International Energy Agency, Paris

IEA (2021) Net zero by 2050 a roadmap for the global energy sector. International Energy Agency, Paris

Iliopoulou EF, Stefanidis SD, Kalogiannis KG, Delimitis A, Lappas AA, Triantafyllidis KS (2012) Catalytic upgrading of biomass pyrolysis vapors using transition metal-modified ZSM-5 zeolite. Appl Catal B 127:281–290

IRENA (2020) Green hydrogen cost reduction: scaling up electrolysers to meet the 1.5 °C climate goal. International Renewable Energy Agency, Abu Dhabi

Jahangiri H, Bennett J, Mahjoubi P, Wilson K, Gu S (2014) A review of advanced catalyst development for Fischer-Tropsch synthesis of hydrocarbons from biomass derived syn-gas. Catal Sci Technol 4(8):2210–2229

Kurkela E, Kurkela M, Hiltunen I (2016) Steam–oxygen gasification of forest residues and bark followed by hot gas filtration and catalytic reforming of tars: results of an extended time test. Fuel Process Technol 141:148–158

Larson ED, Jin H, Celik FE (2009) Large-scale gasification-based coproduction of fuels and electricity from switchgrass. Biofuels, Bioprod Biorefin 3(2):174–194

Liu C, Wang H, Karim AM, Sun J, Wang J (2014) Catalytic fast pyrolysis of lignocellulosic biomass. Chem Soc Rev 43(22):7594–7623

McKendry (2002) Energy production from biomass (part 1): overview of biomass. Bioresour Technol 83(1):37–46

Merrild H, Larsen A, Christensen T (2012) Assessing recycling versus incineration of key materials in municipal waste: the importance of efficient energy recovery and transport distances. Waste Manage 32(5):1009–1018

Miandad R, Barakat MA, Aburiazaiza AS, Rehan M, Nizami AS (2016) Catalytic pyrolysis of plastic waste: a review. Process Saf Environ Prot 102:822–838

Miao Z, Grift TE, Hansen AC, Ting KC (2011) Energy requirement for comminution of biomass in relation to particle physical properties. Ind Crops Prod 33(2):504–513

Milne TA, Evans RJ, Abatzoglou N (1998) Biomass gasifier "tars": their nature, formation, and conversion. National Renewable Energy Laboratory, Report No. NREL/TP-25357

Molino A, Chianese S, Musmarra D (2016) Biomass gasification technology: the state of the art overview. J Energy Chem 25(1):10–25

Obermeier W, Nabel J, Loughran T, Hartung K, Bastos A, Havermann F, Anthoni P, Arneth A, Goll D, Lienert S, Lombardozzi D, Luyssaert S, McGuire, Melton J, Poulter B, Sitch S, Sullivan M, Tian H, Walker A, Wiltshire A, Zaehle S, Pongratz J (2021) Modelled land use and land cover change emissions—a spatio-temporal comparison of different approaches. Earth Syst Dyn 12(2):635–670

Pantaleo A, Richter M, Lombardi P, Arendarski B, Neumann A, Hoepfner A, Komarnicki P (2021) A vision for energy decarbonization: planning sustainable tertiary sites as net-zero energy systems. Energies 14(17):5577–5593

Pérez-Fortes M, Schöneberger JC, Boulamanti A, Tzimas E (2016) Methanol synthesis using captured CO_2 as raw material: techno-economic and environmental assessment. Appl Energy 161:718–732

Peters MS, Timmerhaus KD, West RE (2003) Plant design and economics for chemical engineers, 5th edn. McGraw-Hill, New York

Phillips S, Aden A, Jechura J, Dayton D, Eggeman T (2007) Thermochemical ethanol via indirect gasification and mixed alcohol synthesis of lignocellulosic biomass. Technical Report NREL/TP-510-41168. National Renewable Energy Lab. (NREL), Golden, CO (United States)

Pires A, Martinho G, Chang N (2011) Solid waste management in European countries: a review of systems analysis techniques. J Environ Manage 92(4):1033–1050

Plastics Europe (2019) Plastics—the Facts 2019. An analysis of European plastics production, demand and waste data

Porosoff MD, Yan B, Chen JG (2016) Catalytic reduction of CO_2 by H_2 for synthesis of CO, methanol and hydrocarbons: challenges and opportunities. Energy Environ Sci 9(1):62–73

Pressley PN, Aziz TN, DeCarolis JF, Barlaz MA, He F, Li F, Damgaard A (2014) Municipal solid waste conversion to transportation fuels: a life-cycle estimation of global warming potential and energy consumption. J Clean Prod 70:145–153

Prins MJ, Ptasinski KJ, Janssen FJ (2007) From coal to biomass gasification: comparison of thermodynamic efficiency. Energy 32(7):1248–1259

Ragaert K, Delva L, Van Geem K (2017) Mechanical and chemical recycling of solid plastic waste. Waste Manage 69:24–58

Rentizelas AA, Tolis AJ, Tatsiopoulos IP (2009) Logistics issues of biomass: the storage problem and the multi-biomass supply chain. Renew Sustain Energy Rev 13(4):887–894

Ritchie H, Samborska V, Roser M (2018) Plastic pollution. Published online at OurWorldinData.org. Retrieved from: 'https://ourworldindata.org/plastic-pollution' [Online Resource]

Rubin ES, Davison JE, Herzog HJ (2015) The cost of CO_2 capture and storage. Int J Greenhouse Gas Control 40:378–400

Schulz H (1999) Short history and present trends of Fischer–Tropsch synthesis. Appl Catal 186(1–2):3–12

Searcy EM, Hess JR (2010) Uniform-format feedstock supply system: a commodity-scale design to produce an infrastructure-compatible biocrude from lignocellulosic biomass. Idaho National Laboratory, Idaho Falls, Idaho, INL/EXT-10-20372

Sharuddin SD, Abnisa F, Wan Daud M, Aroua MK (2016) A review on pyrolysis of plastic wastes. Energy Convers Manage 115:308–326

Sikarwar VS, Zhao M, Clough P, Yao J, Zhong X, Memon MZ, Shah N, Anthony EJ, Fennell PS (2016) An overview of advances in biomass gasification. Energy Environ Sci 9(10):2939–2977

Smit E, Weckhuysen BM (2008) The renaissance of iron-based Fischer-Tropsch synthesis: on the multifaceted catalyst deactivation behaviour. Chem Soc Rev 12(37):2758–2781

Swanson RM, Platon A, Satrio JA, Brown RC (2010) Techno-economic analysis of biomass-to-liquids production based on gasification. Fuel 89:S11–S19

Todd TA, Braase LB (2015) Material recovery and waste form development FY 2015 accomplishments report. Idaho National Laboratory

Van der Drift A, Boerrigter H, Coda B, Cieplik MK, Hemmes K (2004) Entrained flow gasification of biomass. Ash behaviour, feeding issues, system analyses. Technical Report. ECN-C-04-039. Energy research Centre of the Netherlands ECN, Petten (Netherlands)

Van der Laan GF, Beenackers AC (1999) Hydrocarbon selectivity model for the gas−solid Fischer−Tropsch synthesis on precipitated iron catalysts. Ind Eng Chem Res 38(4):1277–1290

Vassilev S, Baxter D, Andersen LK, Vassileva CG (2010) An overview of the chemical composition of biomass. Fuel 89(5):913–933

Venderbosch R, Prins W (2010) Fast pyrolysis technology development. Biofuels, Bioprod Biorefin 4(2):178–208

Vervloet D, Kapteijn F, Nijenhuis J, van Ommen JR (2012) Fischer-Tropsch reaction–diffusion in a cobalt catalyst particle: aspects of activity and selectivity for a variable chain growth probability. Catal Sci Technol 2:1221–1233

Wang Y, Dai L, Fan L, Cao L (2017) Catalytic co-pyrolysis of waste vegetable oil and high density polyethylene for hydrocarbon fuel production. Waste Manage 61:276–282

Williams PT (2013) Pyrolysis of waste tyres: a review. Waste Manage 33(8):1714–1728

Woolcock PJ, Brown RC (2013) A review of cleaning technologies for biomass-derived syngas. Biomass Bioenerg 52:54–84

Wright M, Daugaard DE, Satrio JA, Brown RC (2010) Techno-economic analysis of biomass fast pyrolysis to transportation fuels. Fuel 89(S1):S2–S10

Zhou H, Meng A, Long Y, Zhang Y (2015) An overview of characteristics of municipal solid waste fuel in China: physical, chemical composition and heating value. ChemInform (WILEY-VCH Verlag) 46(30)

Glossary of Terms

Artificial Intelligence and LLMs

Attention Mechanism A component in Transformer neural networks that enables models to focus differentially on specific parts of input sequences, allowing the model to weigh the importance of different words when computing representations for each word in the sequence. This mechanism is fundamental to how modern LLMs process context and relationships between words.

Causal Masking A technique used in decoder-only Transformer architectures that ensures each token can only attend to itself and previous tokens during processing, enabling autoregressive text generation. This masking is critical for maintaining the one-directional flow of information during text generation.

Chain of Thought (CoT) A prompt engineering technique that guides LLMs to break down complex problems into sequential reasoning steps, making each step explicit and enhancing reliability for multi-step analysis. This approach significantly improves performance on tasks requiring logical reasoning or procedural understanding.

Context Window The maximum amount of text an LLM can process at once, typically measured in tokens. Current models have finite context windows ranging from 4000–100,000 tokens, constraining the amount of information that can be considered simultaneously.

Embeddings Dense vector representations of words or tokens in a high-dimensional space, where semantically similar tokens are positioned closer together, enabling models to capture meaning relationships. These representations form the foundation of how LLMs understand language.

Few-Shot Learning A technique where LLMs are provided with a small number of examples to establish a pattern for solving similar problems, improving performance on specific tasks without fine-tuning. This approach leverages the model's pattern recognition capabilities.

R. Larraz, *Prompt Engineering and the Transformation of Petroleum Refining*, Synthesis Lectures on Chemical Engineering and Biochemical Engineering,
https://doi.org/10.1007/978-3-031-99728-0

Fine-Tuning The process of further training a pre-trained LLM on specialized datasets with specific objectives to enhance performance for particular domains or tasks. Fine-tuning adapts the general language understanding of the model to specialized contexts.

Hallucination The phenomenon where LLMs generate plausible but factually incorrect information, particularly challenging in technical applications requiring high precision. Mitigating hallucinations is crucial for reliable use in industrial contexts.

Knowledge Cutoff The date after which an LLM does not have information from its training data, requiring explicit updates or external information sources for recent developments. This limitation affects the currency of knowledge available to the model.

Large Language Model (LLM) An artificial intelligence system trained on massive text corpora that can understand, generate, and manipulate human language with remarkable coherence and utility. These models typically contain billions to trillions of parameters.

Logits The raw, unnormalized scores output by a Large Language Model's final linear layer before conversion to probabilities via the softmax function. These values directly represent the model's preference for each token in the vocabulary during text generation. Higher logit values indicate stronger preference for particular tokens, and these values can be manipulated through parameters like temperature to control the randomness of generation. Logits serve as the direct input to the softmax function, which transforms them into a proper probability distribution for token selection.

Meta-prompting An advanced technique where LLMs are used to improve prompt formulation itself, creating increasingly effective instructions based on initial responses or specific requirements. This approach enables continuous improvement of prompting strategies.

Multi-head Attention An extension of the attention mechanism that performs multiple attention operations in parallel, allowing the model to focus on different aspects of relationships between words simultaneously. This enables the capture of diverse linguistic patterns.

One-Shot Prompting A prompt engineering technique that provides a single example to guide the model's response toward solving a specific problem, establishing a pattern for the desired output format or reasoning approach.

Pre-training The initial training phase for LLMs where they learn language patterns from massive datasets through self-supervised learning, typically via next-token prediction. This phase establishes the model's general language understanding capabilities.

Prompt Engineering The systematic development and refinement of inputs to LLMs to elicit desired responses, involving structured approaches to framing problems, directing reasoning processes, and optimizing model outputs for specific applications.

Retrieval Augmented Generation (RAG) A technique that enhances LLM outputs by retrieving relevant information from external knowledge sources before generating responses, addressing knowledge cutoff limitations and improving factual accuracy.

ReAct A prompt engineering approach that integrates Reasoning and Acting steps, enabling LLMs to approach complex problems through an iterative cycle of thought and action. This technique enhances problem-solving capabilities for multi-step challenges.

Role Prompting A technique that directs LLMs to assume a specific professional profile or expertise when generating responses, activating domain-specific knowledge patterns and specialized analytical approaches.

Self-attention The mechanism in Transformer models that allows each token to assess and weigh the relevance of all other tokens in the sequence, enabling context-aware representations. This is the cornerstone of how Transformers process language.

Self-consistency A verification technique that involves generating multiple independent solutions to the same problem, then comparing them to identify areas of consensus or disagreement. This approach enhances reliability for critical applications.

Softmax A mathematical function used in neural networks, particularly in attention mechanisms of Transformer models, that converts a vector of real numbers into a probability distribution. In LLMs, softmax is crucial for converting the model's raw logits into token probabilities during text generation, determining which tokens are most likely to appear next in a sequence. It's formally defined as:

$$\text{softmax}(z)_i = e^{z_i} / \sum_j e^{z_j}$$

where z is a vector of inputs and the output is a vector of values between 0 and 1 that sum to 1.

Tokenization The process of converting text into smaller units (tokens) that can be processed by LLMs, typically using subword tokenization methods like Byte-Pair Encoding. This transforms natural language into a format the model can mathematically process.

Transformer Architecture The neural network design that forms the basis of modern LLMs, characterized by self-attention mechanisms that allow parallel processing of text sequences. This architecture revolutionized natural language processing by enabling more efficient training and better performance.

Tree of Thoughts (ToT) An extension of Chain of Thought that creates a branching structure to explore multiple reasoning pathways simultaneously, comparing their implications for complex decisions. This technique is particularly valuable for problems with multiple viable solution paths.

Zero-Shot Chain of Thought A technique that triggers structured reasoning in LLMs without providing explicit examples, typically using simple phrases like "Let's think step by step" to encourage methodical problem-solving.

Zero-shot Learning The ability of LLMs to perform tasks without specific examples, relying on general knowledge acquired during pre-training. This capability demonstrates the transfer learning potential of large language models.

Petroleum Refining Technology

Alkylation A refining process that combines light olefins with isobutane using acid catalysts to produce high-octane gasoline components with excellent combustion characteristics. This process creates branched-chain paraffins that enhance fuel performance.

Atmospheric Distillation The primary separation process in refineries that separates crude oil into various fractions based on boiling points at atmospheric pressure, creating the foundation for downstream processing. This is typically the first major processing step in a refinery.

Biorefining The processing of biomass to produce fuels, chemicals, and materials, integrating renewable feedstocks into conventional refining operations. This approach represents a key pathway toward more sustainable refining practices.

Carbon Intensity A quantitative measure of greenhouse gas emissions associated with a fuel or process throughout its lifecycle, typically expressed as grams of CO_2 equivalent per unit of energy (g CO_2e/MJ). This metric is fundamental for evaluating the climate impact of different fuels and processes.

Catalytic Cracking A conversion process that uses catalysts (typically zeolites) to break large hydrocarbon molecules into smaller, more valuable products, particularly gasoline. This process significantly increases the yield of high-value products from crude oil.

Catalytic Pyrolysis A thermochemical process that uses catalysts to convert biomass or waste plastics into liquid fuels and chemicals at moderate temperatures (400–600 °C). The catalysts improve product selectivity compared to non-catalytic thermal processes.

Catalytic Reforming A process that converts low-octane naphtha into high-octane gasoline components and aromatic compounds through dehydrogenation, isomerization, and cyclization reactions. This process also produces significant quantities of hydrogen as a by-product.

Coking A thermal cracking process that converts heavy residual oils into lighter products and petroleum coke through carbon rejection. This process handles the heaviest, lowest-value fractions of crude oil.

Fischer–Tropsch Synthesis A catalytic process converting synthesis gas (CO and H_2) into liquid hydrocarbons, central to gas-to-liquids and biomass-to-liquids technologies. This process can produce synthetic fuels with properties similar to conventional petroleum products.

Fluid Catalytic Cracking (FCC) An advanced catalytic cracking technology using fluidized catalyst beds to convert heavy gas oils into gasoline and lighter products with high efficiency. This process is a cornerstone of modern refining for maximizing gasoline production.

Gasification A thermochemical process that converts carbon-containing materials into synthesis gas through partial oxidation under controlled conditions. This versatile process can handle diverse feedstocks from coal to biomass to waste.

Heat Integration Systematic design of process heat exchanger networks to minimize energy consumption by transferring heat between process streams. This approach significantly reduces external energy requirements and improves overall efficiency.

Hydrocracking A catalytic process combining hydrogen with heavy hydrocarbons under high pressure to produce lighter, more valuable products. This process is particularly effective for producing high-quality middle distillates like diesel and jet fuel.

Hydrotreating A process using hydrogen and catalysts to remove impurities (sulfur, nitrogen, metals) from petroleum fractions, improving product quality and reducing emissions. This process is critical for meeting stringent fuel quality specifications.

Isomerization A process that rearranges straight-chain hydrocarbons into branched structures with improved properties without changing the molecular formula. This process is important for enhancing the octane rating of light gasoline components.

Net Zero Refinery An integrated processing facility that produces fuels and chemicals with neutral or negative carbon emissions on a lifecycle basis by using renewable feedstocks, carbon capture technologies, and renewable energy. This concept represents a fundamental transformation of traditional petroleum refining.

Pyrolysis Thermal decomposition of materials at elevated temperatures in the absence of oxygen, used to convert biomass or plastics into bio-oils or chemical feedstocks. This process serves as an initial conversion step for many alternative feedstocks.

Sustainable Aviation Fuel (SAF) Aviation fuel derived from sustainable resources rather than petroleum, offering significantly lower lifecycle carbon emissions while meeting rigorous performance specifications. SAF is critical for decarbonizing the aviation sector.

Syngas A mixture of carbon monoxide and hydrogen produced through gasification, serving as an intermediate for producing fuels and chemicals through catalytic processes. Syngas represents a versatile platform molecule for various downstream conversions.

Technology Readiness Level (TRL) A method for estimating technology maturity on a scale of 1–9, with TRL 1 representing basic principles observed and TRL 9 representing fully commercial technology with extensive operational history. This framework helps assess development stages and implementation risks.

Vacuum Distillation Distillation under reduced pressure to separate heavy fractions without thermal decomposition, enabling recovery of valuable components from atmospheric residue. This process extends the range of petroleum fractions that can be separated by distillation.

Sustainability and Energy Transition

Biogenic Carbon Carbon derived from recently living organisms, considered climate-neutral when released as CO_2 since it was recently captured from the atmosphere through photosynthesis. This concept is central to the carbon accounting of biomass-derived fuels.

Carbon Capture and Storage (CCS) Technology that captures CO_2 emissions from industrial processes and stores them permanently underground to prevent atmospheric release. CCS is critical for mitigating emissions from hard-to-abate industrial sectors.

Carbon Capture, Utilization, and Storage (CCUS) An expansion of CCS that includes converting captured CO_2 into valuable products or using it for enhanced oil recovery, providing economic incentives for capture technologies while reducing net emissions.

Carbon-Negative Processes or products that remove more carbon dioxide from the atmosphere than they emit over their lifecycle, providing a net climate benefit. This concept goes beyond carbon neutrality to actively reduce atmospheric CO_2 concentrations.

Circular Economy An economic system aimed at eliminating waste by continually reusing resources and materials, relevant to refining through waste-to-fuel technologies and recycling of carbon-containing waste streams.

Decarbonization The process of reducing or eliminating carbon dioxide emissions from energy production and industrial processes, central to addressing climate change. This transformation involves fundamental changes to energy systems and industrial processes.

Electrification Replacement of fossil fuel-powered equipment with electrical alternatives, typically coupled with renewable electricity for emissions reduction. Electrification is a key strategy for reducing direct emissions from industrial processes.

Energy Integration Systematic approach to designing processes to minimize energy consumption through heat recovery and process optimization. This approach reduces both energy costs and associated emissions.

Green Hydrogen Hydrogen produced through water electrolysis using renewable electricity, with minimal carbon emissions compared to conventional production methods. Green hydrogen serves as both an energy carrier and chemical feedstock for sustainable systems.

Life Cycle Assessment (LCA) A methodology for assessing environmental impacts associated with all stages of a product's life, from raw material extraction to disposal. LCA provides a comprehensive view of environmental performance beyond direct emissions.

Renewable Diesel A diesel fuel produced from renewable feedstocks that is chemically identical to petroleum diesel, enabling direct use in existing infrastructure without blending limitations. This "drop-in" characteristic facilitates adoption.

Industrial Design and Implementation

Constraint-based Design A design approach that explicitly incorporates physical, operational, economic, and regulatory constraints to ensure practical solutions. This methodology ensures that designs remain feasible within real-world limitations.

First Principles Reasoning An approach to problem-solving that relies on basic, foundational concepts rather than analogies or previous examples. This approach enables novel solutions by examining fundamental scientific and engineering principles.

Hierarchical Decomposition A method that breaks complex systems into manageable components while maintaining awareness of their interconnections. This approach facilitates analysis of complex systems without losing sight of integration requirements.

Integration-focused Design A design approach emphasizing connections and interfaces between components rather than optimizing components in isolation. This methodology enhances overall system performance through synergistic interactions.

Multi-perspective Analysis A technique for analyzing problems from diverse stakeholder perspectives to ensure comprehensive understanding before solution development. This approach identifies requirements and constraints that might otherwise be overlooked.

Phased Implementation A strategic approach to project execution that divides implementation into sequential phases with defined milestones and decision points. This approach manages risk while enabling progressive learning and capability building.

Risk-opportunity Mapping Systematic analysis identifying both challenges requiring mitigation and potential advantages that could be leveraged during implementation. This balanced view enhances strategic planning for complex projects.

Scenario-based Exploration A technique that tests potential solutions against different possible futures to develop approaches that remain robust across varying circumstances. This methodology enhances resilience to uncertainty.

Trade-off Elicitation Method for explicitly mapping inherent compromises between competing objectives in design decisions, facilitating informed prioritization. This approach acknowledges that most complex decisions involve balancing multiple factors.

Transformation The fundamental redesign and reimagining of industrial systems to achieve step-change improvements in performance, sustainability, or other key metrics. This process goes beyond incremental improvement to fundamentally alter how systems operate.

MIX
Papier aus verantwortungsvollen Quellen
Paper from responsible sources
FSC® C105338

If you have any concerns about our products,
you can contact us on
ProductSafety@springernature.com

In case Publisher is established outside the EU,
the EU authorized representative is:
Springer Nature Customer Service Center GmbH
Europaplatz 3, 69115 Heidelberg, Germany

Printed by Libri Plureos GmbH
in Hamburg, Germany